The History of Science

THE HISTORY OF
MATHEMATICS

The History of Science

THE HISTORY OF
MATHEMATICS

Anne Rooney

ROSEN
PUBLISHING®

New York

Acknowledgments

*With thanks to those of my Facebook friends who have helped in
various ways, particularly Michael Anti [Zhao Jing] (Harvard
Faculty/Cambridge Faculty/Boston MA), Gordon Joly (London),
John Naughton (Cambridge Alum '68, The Open University
Faculty), Jack Schofield (London/Guardian News and Media), and
Bill Thompson (London/Cambridge Faculty/City UK Faculty).*

Additional thanks to Professor Robin Wilson, The Open University.

This edition published in 2013 by:

The Rosen Publishing Group, Inc.
29 East 21st Street, New York, NY 10010

Library of Congress Cataloging-in-Publication Data

Rooney, Anne.
  The history of mathematics / Anne Rooney. -- 1st ed.
      p. cm. -- (The history of science)
  Includes bibliographical references and index.
  ISBN 978-1-4488-7227-5 (library binding)
  1. Mathematics--Philosophy. 2. Mathematics--History. I. Title.
  QA8.4.R66 2013
  510.9--dc23
                        2012009972

Manufactured in China

SL002276US

CPSIA Compliance Information: Batch #S12YA: For further information, contact
Rosen Publishing, New York, New York, at 1-800-237-9932

Contents

THE MAGIC OF NUMBERS

Think of a number from 1 to 9.
Multiply it by 9.
If you have a two-digit number,
add the digits together.
Take away 5.
Multiply the number by itself.

The answer is 16. How does it work?
It all depends on a crucial bit of number
magic: adding together the digits of
multiples of 9 always gives 9:

  9: 0 + 9 = 9
  18: 1 + 8 = 9
  27: 2 + 7 = 9 and so on.

Thereafter, it's all plain sailing:

  9-5 = 4, 4 × 4 = 16.

There is plenty more magic in numbers. Long ago, some of the earliest human civilizations discovered the strange and fascinating quality of some numbers and wove them into their superstitions and religions. Numbers have entranced people ever since, and still hold the power to unlock the universe for us, by providing a key to the secrets of science. Our understanding of everything from the behavior of subatomic particles to the expansion of the universe is based on mathematics.

MATH FROM THE START

The earliest records of mathematical activity (beyond counting) date from 4,000 years ago. They come from the fertile deltas of the Nile (Egypt) and the plains between the Tigris and Euphrates (Mesopotamia, now Iraq). We know little of the individual mathematicians of these early cultures.

Around 600 BCE the Ancient Greeks developed an interest in mathematics. They went beyond their predecessors in that they were interested in finding rules that could be applied to any problem of a similar type. They worked on concepts in mathematics that underlie all that has come since. Some of the greatest mathematicians of all time lived in Greece and the Hellenic center of Alexandria in Egypt.

As the Greek civilization came to an end, mathematics in the West entered a dead zone. Several hundred years later, Islamic scholars in the Middle East picked up the baton. Baghdad, built around 750, became a dazzling intellectual center where Arab Muslim scholars pulled together the legacy

Toledo in Spain became the gateway through which Arab learning entered Europe in the late 11th century.

of both Greek and Indian mathematicians and forged something new and dynamic. Their progress was greatly aided by their adoption of the Hindu-Arabic number system that we now use, and given impetus by their interests in astronomy and optics, as well as the requirements of the Islamic calendar and the need to find the direction of Mecca. However, the demands of Islam, which were once a spur to development, eventually stifled further growth. Muslim theology ruled against intellectual activity that was considered spiritually dangerous—in that it might uncover truths that should stay hidden, or challenge the central mysteries of religion.

Luckily, the Arab presence in Spain made the transfer of mathematical knowledge to Europe quite straightforward. From the late 11th century, Arab and Greek texts were translated into Latin and spread rapidly around Europe.

There was little new development in mathematics in Europe during the Middle Ages. At the point where a few people were equipped to carry mathematics forward, Europe was struck by the cataclysm of the Black Death (1347–50). Between a quarter and a half of the population died in many European countries. It was the 16th century before much new progress was made, but then there was a flurry of intellectual activity, in mathematics as in science, art, philosophy, and music. The invention in Europe of the printing press accelerated the spread of new learning. European mathematicians and scientists began to shape modern mathematics and to find myriad applications for it.

While this has been the path of development of present-day mathematics, many cultures have developed in parallel, often making identical or comparable discoveries but not feeding into the main story centered on North Africa, the Middle East, and Europe. China kept itself separate from the rest of the world for thousands of years, and Chinese mathematics flourished independently. The meso-American societies in South America developed their own mathematical systems too, but they were wiped out by European invaders and colonists who arrived in the 16th century. Early Indian mathematics did feed into the Arab tradition from around the ninth century, and in recent years India has become a rich source of world-class mathematicians.

At the very end of our story, a single number system and mathematical ethos has spread around the globe, and mathematicians from all cultures including Japan, India, Russia, and the United States work alongside those of Europe and the Middle East toward similar goals. Though mathematics is now a global enterprise, it has only recently become so.

CHAPTER 1

STARTING
with Numbers

Before we could have mathematics, we needed numbers. Philosophers have argued for years about the status of numbers, about whether they have any real existence outside human culture, just as they argue about whether mathematics is invented or discovered. For example, is there a sense in which the area of a rectangle "is" the multiple of two sides—which is true, independent of the activity of mathematicians? Or is the whole a construct, useful in making sense of the world as we experience it, but not "true" in any wider sense? The German mathematician Leopold Kronecker (1823–91) made many enemies when he wrote, "God made the integers; all else is the work of man." Whichever opinion we incline toward individually, it is with the positive integers—the whole numbers above zero—that humankind's mathematical journey began.

In the beginning... cavemen could paint, but could they count?

We regulate all aspects of our life by numbers, but that has not always been the case. The minute hand was added to clocks in 1475, the second hand around 1560.

Where Do Numbers Come From?

Numbers are so much a part of our everyday lives that we take them for granted. They're probably the first thing you see in the morning as you glance at the clock, and we all face a barrage of numbers throughout the day. But there was a time before number systems and counting. The discovery—or invention—of numbers was one of the crucial steps in the cultural and civil development of humankind. It enabled ownership, trade, science, and art, as well as the development of social structures and hierarchies—and, of course, games, puzzles, sports, gambling, insurance, and even birthday parties!

CAN ANIMALS COUNT?

Could the mammoths count their attackers? Some animals can apparently count small numbers. Pigeons, magpies, rats, and monkeys have all been shown to be able to count small quantities and distinguish approximately between larger quantities. Many animals can recognize if one of their young is missing, too.

Four Mammoths or More Mammoths?

Imagine an early human looking at a herd of potential lunch—buffalo, perhaps, or woolly mammoths. There are a lot; the hunter has no number system and can't count them. He or she has a sense of whether it is a large herd or a small herd,

Many against one is more likely to ensure a safe outcome and a meal for hunters equipped only with primitive weapons.

HOW TO COUNT SHEEP
WITHOUT COUNTING

As each sheep leaves the pen, make a notch on a bone or put a pebble in a pile. When it's sheep bedtime, check a notch or a pebble for each sheep that comes in.

- If there are pebbles or notches unaccounted for, go and look for the lost sheep.
- If a sheep dies, lose a pebble or scratch out a notch.
- If a sheep gives birth, add a pebble or a notch.

recognizes that a single mammoth makes easier prey, and knows that if there are more hunters the task of hunting is easier and safer. There is a clear difference between one and "more-than-one," and between many and few. But this is not counting.

At some point, it becomes useful to quantify the extra mammoths in some way—or the extra people needed to hunt them. Precise numbers are still not absolutely essential, unless the hunters want to compare their prowess.

TALLY-HO!

Moving on, and the mammoth hunters settle to herding their own animals. As soon as people started to keep animals, they needed a way to keep track of them, to check whether all the sheep/goats/yaks/pigs were safely in the pen. The easiest way to do this is to match each animal to a mark or a stone, using a *tally*.

It isn't necessary to count to know whether a set of objects is complete. We can glance at a table with 100 places set and see instantly whether there are any places without diners. One-to-one correspondence is learned early by children, who play games matching pegs to holes, toy bears to beds, and so on, and was learned early by humankind. This is the basis of set theory—that one group of objects can be compared with another. We can deal simply with sets like this without a concept of number. So the early farmer can move pebbles from one pile to another without counting them.

The need to record numbers of objects led to the first mark-making, the precursor of writing. A wolf bone found in the Czech Republic carved with notches more than 30,000 years ago apparently represents a tally and is the oldest known mathematical object.

FROM TWO TO TWO-NESS

A tally stick (or pile of pebbles) that has been developed for counting sheep can be put to other uses. If there are thirty sheep-tokens, they can also be used for tallying thirty goats or thirty fish or thirty days. It's likely that tallies were used early on to count time—moons or days until the birth of a baby, for example, or from planting to cropping. The realization that "thirty" is a transferable idea and has some kind of independence of the concrete objects counted heralds a concept of number. Besides seeing that four apples can be shared out as two apples for each of two people, people discovered that four of anything can always be divided into two groups of two and, indeed, four "is" two twos.

At this point, counting became more than tallying and numbers needed names.

ONE, TWO, A LOT

A tribe in Brazil, the Pirahã, have words for only "one," "two," and "many." Scientists have found that not having words for numbers limits the tribe's concept of numbers. In an experiment, they discovered that the Pirahã could copy patterns of one, two, or three objects, but made mistakes when asked to deal with four or more objects. Some philosophers consider it the strongest evidence yet for linguistic determinism—the theory that understanding is ring-fenced by language and that, in some areas at least, we can't think about things we don't have words for.

How many have we got? A Portuguese vineyard worker notches a tallystick to record each basket of grapes that passes by.

BODY COUNTING

Many cultures developed methods of counting by using parts of the body. They indicated different numbers by pointing at body parts or distances on the body following an established sequence. Eventually, the names of the body parts probably came to stand for the numbers and "from nose to big toe" would mean (say) 34. The body part could be used to denote 34 sheep, or 34 trees, or 34 of anything else.

TOWARD A NUMBER SYSTEM

Making a single mark for a single counted object on a stick, slate, or cave wall is all very well for a small number of objects, but it quickly becomes unmanageable. Before humankind could use numbers in any more complex way than simply tallying or counting, we needed methods of recording them that were easier to apprehend at a glance than a row of strokes or dots. While we can only surmise from observing non-industrialized people as to how verbal counting systems may have developed, there is physical evidence in the form of artifacts and records for the development of written number systems.

The earliest number systems were related to tallies in that they began with a series of marks corresponding one-to-one to counted objects, so "III" or "..." might represent 3. By 3400 BCE, the Ancient Egyptians had developed a system of symbols (or hieroglyphs) for powers of ten, so that they used a stroke for each unit and a symbol for 10, then a different symbol for 100, another for 1,000, and so on up to 1,000,000. Within each group, the symbol was repeated up to nine times, grouped in a consistent pattern to make the number easy to recognize. In Mesopotamia (current-day Iraq), a similar system existed from at least 3000 BCE.

A still-familiar simple grouping system is Roman numerals. Numbers 1 to 4 are represented by vertical strokes:

I, II, III, IIII

The Romans gave up at IIII, switching to a symbol for five, V. Later, they sometimes used IV for IIII. In this case the position of the vertical stroke determines its meaning—five minus one. In the same way, IX is used for nine (ten minus one).

Different symbols are used to denote multiples of five and ten:

V = 5	L = 50	D = 500
X = 10	C= 100	M = 1,000

Numbers are built up by grouping units, tens, and so on. So 2008 is represented by MMVIII. The characters for 5, 50, and 500 can't be used more than once in a number, since VV is represented by X, and so on.

Some numbers are quite laborious to write. For example, 38 is written XXXVIII. The system doesn't allow subtraction from anything except the next symbol in the numerical sequence, so 49 can't be written IL (50 minus 1); it has to be written XLIX (50 minus 10; 10 minus 1).

The next step is a system that instead of repeating the symbols for a number (XXX for 30, for instance) uses a symbol for each of the digits

Early Egyptian hieroglyphs represented numbers using powers of ten, and could show numbers up to 9,999,999.

1 to 9, and then this is used with the symbols for 10, 100, and so on to show how many 10s, 100s, and 1,000s are intended. The current Chinese system works on this principle. So:

四十 = 4 x 10 = 40

but 十四   10 + 4 = 14

and 四十四   4 × 10 + 4 = 44

This is known as a multiplicative grouping system. The number of characters needed to represent numbers is more regular with this type of system. Numbers 1 to 10 are shown by one digit; numbers 11 to 20 are shown by two digits; thereafter, multiples of 10 up to 90 are shown by two digits (20, 30, etc.) and the other numbers up to 99 are

shown by three digits. Roman numerals, on the other hand, need between one and four digits for the numbers 1 to 10 and between one and eight digits for numbers up to 100.

CIPHERED SYSTEMS

The hieroglyphic system described above (see page 13) was only one of three systems used in Ancient Egypt. There were two ciphered systems, demotic and hieratic. A ciphered system not only has different symbols for the numerals 1 to 9, but distinct symbols for each of the multiples of 10, 100, and 1,000. Hieratic is the oldest known ciphered system. It could express numbers in a very compact form, but to use it people must learn a large number of different symbols. This may have served a social purpose, keeping numbers "special" and so endowing those who knew them

HOW OLD IS THE COW AND HAVE YOU BEEN PAID?

In Babylon (from southern Iraq to the Persian Gulf), two systems of writing numbers were used. One, cuneiform, consists of wedge-shaped marks made by a stylus in damp clay which was then baked. A different system, curvilinear, was made using the other end of the stylus, which was round. The two scripts were used to represent numbers for different purposes. Cuneiform was used to show the number of the year, the age of an animal, and wages due. Curvilinear was used to show wages that had already been paid.

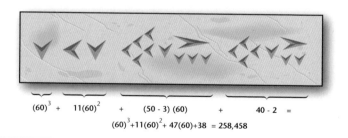

$$(60)^3 + \quad 11(60)^2 \quad + \quad (50-3)(60) \quad + \quad 40-2 \quad =$$
$$(60)^3 + 11(60)^2 + 47(60) + 38 = 258,458$$

Egyptian hieratic numerals of the New Kingdom (1600–1000 BCE) used more symbols than before, making numbers more compact but harder to learn to use.

with extra power, forming a mathematical elite. In many cultures, numbers have been closely allied with divinity and magic, and preserving the mystery of numbers helped to maintain the authority of the priesthood. Even the Catholic Church was to indulge in this jealous guardianship of numbers in the European Middle Ages. Other ciphered systems include Coptic, Hindu Brahmin, Hebrew, Syrian, and early Arabic. Ciphered systems often use letters of the alphabet to represent numerals.

GETTING INTO POSITION

Positional number systems, such as our own modern system, depend on the position of a digit to show its meaning. A positional system develops from a multiplicative grouping system such as the Chinese one by omitting the characters that represent 10, 100, and so on and depending only on the position of the numerals to show their meaning. This can only work when there is a symbol for zero, as otherwise there is no way of distinguishing between numbers such as 24, 204, and 240, a problem encountered by the Babylonians.

	10,000	1,000	100	10	1
54,321 =	$5 \times 10,000$	$4 \times 1,000$	3×100	2×10	1×1
10,070 =	$1 \times 10,000$	$0 \times 1,000$	0×100	7×10	1×0

A positional system can show very large numbers as it does not need new names or symbols each time a new power of 10 is reached.

The earliest positional system that can be dated was developed by the Sumerians from 3000 to 2000 BCE, but it was a complicated system that used both 10 and 60 as its bases. It had no zero until the third century BCE, leading to ambiguity and probably confusion. Even after zero was introduced, it was never used at the end of numbers, so it was only possible to distinguish between, say, 2 and 200

15

SUMERIANS AND BABYLONIANS

The fertile area of Mesopotamia, between the rivers Tigris and Euphrates, has been called the cradle of civilization. Now in Iraq, it was settled by the Sumerians, who by the middle of the fourth millennium BCE had established perhaps the earliest civilization in the world. Invading Akkadians in the 23rd century BCE largely adopted Sumerian culture. The period from around 2000 BCE to 600 BCE is generally called Babylonian. After this, Persian invaders took over, but again continued rather than replaced the culture of the area.

from the context. This was sometimes easy and sometimes not. The statement "I have 7 sons" was unlikely to be interpreted as "I have 70 sons"—but a statement such as "An army of 3 is approaching" contains dangerous ambiguity. An army of 300? No problem. An army of 3,000, or 30,000, or even 300,000 is a very different matter.

One of the two number systems in use in Ancient Greece, that most popular in Athens, used letters of the Greek alphabet to represent numbers, beginning with *alpha* for 1, *beta* for 2, and so on up to 9. Next, individual letters were used for multiples of ten and then for multiples of 100, so that any three-digit number could be represented by three letters, any four-digit number by four letters, and so on. They didn't have enough letters in their alphabet to make it up to 900 with this system, so some of the numerals were represented by

archaic letters they no longer used for writing. For numbers over 999 they added a tick mark to the right of a letter to show that it must be multiplied by a factor of 1,000 (like our comma as a separator) or the letter *mu* as a subscript to show multiplication by 10,000. To distinguish numbers from words, they drew a bar over numbers. Greek philosophers later came up with methods of writing very large numbers, not because they especially needed them, but to counter claims that larger numbers could not exist since there was no way of representing them.

The Mayans used a complete positional system, with a zero, used thoroughly. The earliest known use of zero in a Mayan inscription is 36 BCE. Mayan culture was discovered—and consequently wiped out, along with the Mayan civilization—by Spanish invaders who came to Yucatan in the early 16th century. The Mayan number system was based on 5 and 20 rather than 10, and again had limitations. The first perfect positional system was the work of the Hindus, who used a dot to represent a vacant position.

THE BIRTH OF HINDU-ARABIC NUMBERS

The numbers we use today in the West have a long history and originated with the Indus valley civilizations more than 2,000 years ago. They are first found in early Buddhist inscriptions.

The use of a single stroke to stand for "one" is intuitive and, not surprisingly, many cultures came up with the idea. The orientation of the stroke varies—while in the West we still use the Hindu-Arabic vertical stroke (1), the Chinese use a horizontal stroke (–). But what about the other numbers? The squiggles we now use to represent 2, 3, 4, and so on?

The earliest, 1, 4, and 6, date from at least the third century BCE and are found in the Indian Ashoka inscriptions (these record thoughts and deeds of the Buddhist Mauryan ruler of India, Ashoka the Great, 304–232 BCE). The Nana Ghat inscriptions of the second century BCE added 2, 7, and 9 to the list, and 3 and 5 are found in the Nasik caves of the first or second century CE. A text written by the Christian Nestorian bishop Severus Sebokht

living in Mesopotamia about 650 CE refers to nine Hindu numbers.

1	2	3	4	5	6	7	8	9
–	=	≡	+	♭	⏀	?	↰	?

Adding a diagonal line between the horizontal strokes of the Brahmi "2" and a vertical line to the right of the strokes of the Brahmi "3" makes recognizable versions of our numerals.

The Brahmi numerals were part of a ciphered system, with separate symbols for 10, 20, 30, and so on.

MOVING WESTWARD

The Arab writer Ibn al-Qifti (1172–1242) records in his *Chronology of the Scholars* how an Indian scholar brought a book to the second Abisid Caliph Abu Ja'far Abdallah ibn Muhammad al-Mansur (712–75) in Baghdad, Iraq, in 766. The book was

BRAHMAGUPTA (589–668)

The Indian mathematician and astronomer Brahmagupta was born in Bhinmal in Rajasthan, northern India. He headed the astronomical observatory at Ujjain and published two texts on mathematics and astronomy. His work introduced zero and rules for its use in arithmetic, and provided a way of solving quadratic equations equivalent to the formula still used today:

$$\frac{-b \pm \sqrt{b^2 - 4ac}}{2a}$$

Brahmagupta's text *Brahmasphutasiddhanta* was used to explain the Indian arithmetic needed for astronomy at the House of Wisdom.

probably the *Brahmasphutasiddhanta* (The Opening of the Universe) written by the Indian mathematician Brahmagupta in 628. The caliph had founded the House of Wisdom, an educational institute that led intellectual development in the Middle East at the time, translating Hindi and Classical Greek texts into Arabic. Here, the *Brahmasphutasiddhanta* was translated into Arabic and Hindu numbers took their first step toward the West.

The diffusion of the Indian numerals throughout the Middle East was assured by two very important texts produced at the House of Wisdom: *On the Calculation with Hindu Numerals* by the Persian mathematician al-Khwarizmi (*ca.* 825), and *On the Use of the Indian Numerals* by the Arab Abu Yusuf Yaqub ibn Ishaq al-Kindi (830).

A system of counting angles was adopted for depicting the numerals 1 to 9. It's easy to see how the Hindu numerals could be converted by the addition of joining lines to fit this system—try counting the angles in the straight-line forms of the numerals we use now:

123456789

MUHAMMAD IBN MUSA AL-KHWARIZMI, CA.780–850

The Persian mathematician and astronomer al-Khwarizmi was born in Khwarizm, now Khiva in Uzbekistan, and worked at the House of Wisdom in Baghdad. He translated Hindu texts into Arabic and was responsible for the introduction of Hindu numerals into Arab mathematics. His work was later translated into Latin, giving Europe not just the numerals and arithmetic methods but also the word "algorithm" derived from his name.

When al-Khwarizmi's work was translated, people assumed that he had originated the new number system he promoted and it became known as "algorism." The algorists were those who used the Hindu-Arabic positional system. They were in conflict with the abacists, who used the system based on Roman numerals and calculated with an abacus.

Zero was adopted around the same time; zero, of course, has no angles. The Arab scholars devised the full positional system we use now, abandoning the ciphers for multiples of ten used by the Indian mathematicians.

Not long after, the new fusion of Hindu-Arabic number systems made its way to Europe through Spain, which was under Arab rule. The earliest European text to show the Hindu-Arabic numerals was produced in Spain in 976.

ROMANS OUT!

Of course, Europe was already using a number system when the Hindu-Arabic notation arrived in Moorish Spain. After the fall of the Roman Empire in the West, traditionally dated 476 CE, Roman culture was only slowly eroded.

The Roman number system was unchallenged for over 500 years. Although the Hindu-Arabic numerals crop up in a few works produced or copied in the tenth century, they did not enter the mainstream for a long time.

1	I	5,000	Ⅽ
5	V	10,000	Ⅽ
10	X	50,000	Ⅽ
50	L	100,000	Ⅽ

A FUSS ABOUT NOTHING

The concept of zero might seem the antithesis of counting. While zero was only an absence of items counted, it didn't need its own symbol. But it did need a symbol when positional number systems emerged. Initially, a space or a dot was used to indicate that no figure occupied a place; the earliest preserved use of this is from the mid-second millennium BCE in Babylon.

The Mayans had a zero, represented by the shell glyph:

This was used from at least 36 BCE, but had no influence on mathematics in the Old World. It may be that Meso-Americans were the first people to use a form of zero.

Zero came to the modern world from India. The oldest known text to use zero is the Jain *Lokavibhaaga,* dated 458 CE. Brahmagupta wrote rules for working with zero in arithmetic in his *Brahmasphutasiddhanta,* setting out, for instance, that a number multiplied by zero gives zero. This is the earliest known text to treat zero as a number in its own right.

Al-Khwarizmi introduced zero to the Arab world. The modern name, "zero," comes from the Arab word *zephirum* by way of Venetian (the language spoken in Venice, Italy). The Venetian mathematician Luca Pacioli (1445–1514 or 1517) produced the first European text to use zero properly.

While historians do not count a "year zero" between the years 1 BCE and 1 CE, astronomers generally do.

LETTERS FROM ABROAD

The Romans used written numerals before they could read and write language. They adopted numbers from the Etruscans, who ruled Rome for around 150 years. When the Romans later conquered the Greek-speaking city of Cumae, they learned to read and write. They then adapted the numerals they had taken from the Etruscans to make Roman letters.

As the Empire grew in extent and sophistication, the Romans needed larger and larger numbers. They developed a system of enclosing figures in a box, or three sides of a box, to show that they should be multiplied by 1,000 or 100,000. The system wasn't used consistently, though, so

$\overline{|V|}$ could mean either 5,000 or 500,000.

Arithmetic is virtually impossible with Roman numerals and this was to lead to its eventual replacement.

$$\begin{array}{r} \text{XXXVIII} + \\ \underline{\text{XIX}} \\ \text{LVII} \quad (38 + 19 = 57) \end{array}$$

For the purposes of accounting, taxation, census taking, and so on, Roman accountants always used an abacus. Hindu-Arabic numerals offered a considerable advantage in that the positional system made arithmetic with written numbers very easy. Both Fibonacci (Leonardo Pisano, *ca*.1170–1250) and Luca Pacioli, both better known for other achievements, were instrumental in popularizing the

Fibonacci, the Italian mathematician, learned about Hindu-Arabic numerals as a boy while traveling in North Africa with his trader father.

Hindu-Arabic system, particularly among the merchants and accountants. Even so, it took many centuries and considerable struggle before Europe moved over completely to the use of the Hindu-Arabic system (see Unspeakable Numbers, page 56).

Roman numerals continued to be used for many things long after they were replaced in mathematical functions. They

"The nine Indian figures are:
9 8 7 6 5 4 3 2 1
With these nine figures, and with the sign 0… any number may be written."
Fibonacci, *Liber Abaci*, 1202

NOT OVER YET

It would be a mistake to think that our numbers have stopped evolving. In the last century we have seen the development and subsequent decline of the zero with a slash through it, Ø, to distinguish it from capital "O" in computer printouts, and the representation of digits as a collection of straight lines so that they can be shown by illuminating bars on an LED display. Computer-readable character sets, too, have been developed for use on checks and other financial documents, taking our numerals far from their cursive origins.

are still often used on clock faces, for example, and to show the copyright date of movies and some TV programs.

In addition, we have developed new types of notation for writing numbers so unimaginably large that our ancestors could have had no conceivable use for them (see pages 26–33).

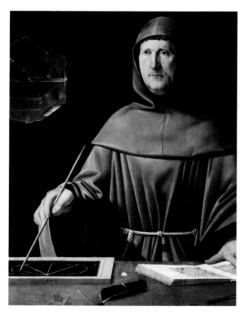

Luca Pacioli was a Franciscan friar. In this portrait by Jacopo de' Barbari (ca.1495), he is demonstrating one of Euclid's theorems.

Bar codes use lines of different thicknesses to represent numbers: these are read by computerized scanners, which "see" them as numbers.

Numbers and Bases

We use a base-10 number system, called a decimal system. In fact, though, whatever system we used we would call it "base 10" because that's how we define "base"—it's the point at which we give up counting units and start counting groups. In a positional number system, it means that we start reusing digits with "10" meaning "all the digits we have, plus 1." In a base-2 system, 2 is represented by 10, and in a base-5 system, 5 is represented by 10. For us, "9+1" is represented by 10.

FINGERS AND THUMBS

We have probably developed our decimal system because most people have ten fingers. Although it seems very obvious that we can count on our fingers, different cultures through the ages have developed different ways of doing it. Fingers may be extended or folded down to indicate a number; joints may be counted as well as fingers; one hand may be used to show tens and the other units, or interaction between people may be required— taking hold of or pulling on the fingers, for example.

A highly developed system, more complicated than ordinary finger counting, was used in both Europe and the Middle East. It was more like a sign language, and enabled counting up to 10,000 and even beyond by making shapes with the fingers. It was evidently in use for a long time as it is described by the seventh-century English writer the Venerable Bede (ca.672–735) and in a 16th-century Persian dictionary, *Farhangi Djahangiri*.

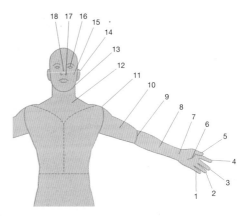

People don't just count on fingers. This body-counting system is used by the Fasu of Papua New Guinea.

BACK TO BASE

Despite the obvious recourse to fingers as a counting aid, not all cultures have used a decimal system of counting. Indeed, we owe many of our strange weights and

ORIENTAL FINGER BARGAINING

Secret systems of bargaining with the fingers were widespread between Algeria and China for centuries. The two participants needed to know the approximate price they were negotiating—whether it would be in units, tens, hundreds, or thousands. One negotiator would hold the index finger of the other to indicate 1 (or 10, or 100), the index and middle fingers to indicate 2, and so on. Clasping the whole hand meant 5. In different places, different methods were used for the numbers above 5. In some places, for instance, 6 was indicated by twice gripping the fingers for 3, in others by grasping the thumb and little finger.

The international language of commerce: a tourist bargains with a local craft vendor near the site of the Terracotta Army, Xian, China.

Often, the negotiators hid their hands inside a sleeve or concealed them in a robe so that onlookers could not see the price agreed.

measures to cultures that have used different counting systems.

Binary, or base 2, is used by computers as it can designate one of two states, TRUE/FALSE, or hold a negative or positive electrical charge. But there have been human users of binary systems. Some of the oldest tribes in Australia use a counting system in which the names of the numbers are defined in relation to two and one. The Gapapaiwa of Milne Bay have *sago* "one," *rua* "two," then *rua ma sago* for "three" (literally, "two and one"), *rua ma rua* or "two and two" for "four," and *rua ma rua ma sago* (two and two and one) for "five." Although it differs from computer binary in that it uses one and two rather than zero and one, it still has only two distinct numbers.

The indigenous peoples of Tierra del Fuego and parts of South America have used number systems with bases 3 and 4. Base-4 systems may have emerged because four is the largest number of items in a row that most people can intuitively apprehend without counting. For this reason, the "five–barred gate" method of tallying has been widely used for counting everything from sheep in a field to days spent in prison.

卌 卌 卌 ||

= 5 + 5 + 5 + 2 = 17

This "rule of four" lies behind many cultural oddities. In ancient Rome, for

There are many ways of counting on the fingers.
The most common use base 5 or base 10.

instance, the first four sons were given "proper" names, such as Marcus or Julius, but after four they were given number-names: Quintus (fifth), Sextus (sixth), Septimus (seventh), and so on.

A few cultures use a quinary (base-5) system, including the speakers of Saraveca, a South American Arawakan language, the Ilongot, a head-hunting tribe from the Philippines, and some Indonesian societies. The Incas had a quinary system with names for numbers up to at least 10,000. It is easy to see how a quinary system could have evolved from counting on the fingers of one hand.

Other common systems are base 6, duodecimal (base 12), and vigesimal (base 20). Bases 12 and 20 have often been used with other bases in a complex system where a small base number is used for low numbers (up to 5 or 10) and a large one for numbers over a certain limit. Remnants of a base-20 system linger on in the French "quatre-vingts" for 80, for example. Vestiges of a base-12 system are all around

us, in the widely used dozen and gross (144 = 12 × 12), the 12 inches in a foot, and 12 months in a year.

The ancient Sumerians had a sexagesimal system—one that used base 60. It is clearly difficult to remember 60 different names for the digits, so for lower numbers they used base 10. From the Babylonians, we still have 60 seconds in a minute and 60 minutes in an hour; when we write 2 hrs 14 mins 38 secs we are using their base-60 system. It is not entirely clear why they used 60, but sixty has many factors (numbers it can be divided by), making it a useful base. Arab mathematicians used base 60 in astronomy, usually converting to decimal for difficult calculations and then moving back to 60 to express their final result.

HOW MANY FINGERS DOES A COMPUTER HAVE?

For bases higher than 10, we need to rope in other symbols to stand for digits we don't have in our decimal system. Computers count either in binary or in hexadecimal (base 16 = 2^4). To represent the numbers between 9 and 10 (= 16) in hexadecimal notation we use letters of the alphabet.

Decimal equivalent	1	2	3	4	5	6	7
Base 2 (binary)	1	10	11	100	101	110	111
Base 3	1	2	10	11	12	20	21
Base 4	1	2	3	10	11	12	13
Base 5	1	2	3	4	10	11	12
Base 6	1	2	3	4	5	10	11

Decimal	9	10	11	12	13	14	15
Hexadecimal	9	A	B	C	D	E	F
Decimal	16	17	18	19	20	21	22
Hexadecimal	10	11	12	13	14	15	16
Decimal	23	24	25	26	27	28	29
Hexadecimal	17	18	19	1A	1B	1C	1D
Decimal	30	31	32	33	34	35	36
Hexadecimal	1E	1F	20	21	22	23	24

Clearly all numbers above 10 mean different things depending on the base system used, so there is plenty of scope for confusion: "11" in hexadecimal means 17 in decimal. Computer books often use the hash sign, #, before a hexadecimal number, so "#11" = 17 (and 23 would be represented by #17).

Because computers count in hexadecimal, some strange numbers are beginning to creep back into everyday life, too. While we still buy eggs by the dozen, we might also buy a memory card that will hold 512 MB of data, or an iPod with 8 GB of storage. The decimal system has by no means taken over completely.

CARTOON COUNTING

Cartoon characters are most often drawn with three fingers and a thumb. Had we all evolved to look like Homer Simpson (though perhaps less yellow, slimmer, and with more hair), we might now use a base-8 counting system in which "10" donuts would be only 8 donuts.

Many items are traded in quantities that don't relate to the decimal system—such as eggs sold by the dozen.

More Numbers, Big and Small

The first numbers humankind used related directly to objects in the real world; they were the positive integers. But these are not the only numbers we recognize now. As time has passed, we have developed ways of quantifying an absence with negative numbers, showing fragments or portions that are less than one, and representing numbers so large they tax our normal systems for writing numbers. We have even developed a way of talking about imaginary (complex) numbers.

INTEGERS

Integers are the positive and negative whole numbers, extending infinitely in both directions from (and including) zero. The positive integers are called natural numbers. Natural numbers have a special status because they can be related one-to-one to indivisible objects in the real world.

For convenience, infinity is the name given to the largest possible number, the end of counting—though clearly this could never exist since however large a number is we could always add one more, and another, and so on. Infinity is represented by the symbol ∞, first used by John Wallis in his book *De sectionibus conicis* (Of conical sections), published in 1655.

LESS THAN ZERO

Negative numbers don't relate directly to the physical world in that we can't count a negative number of objects—we can't see

The scale on a thermometer spans negative and positive temperatures, and is a familiar application of negative numbers.

> *"God created the integers. All the rest is the work of Man."*
> Leopold Kronecker, 1823–91

"minus two cows," for example. But as soon as concepts of ownership emerge, negative numbers have a meaning. They were used early on to indicate a debt (money or goods owed). They are also used in some types of scaled measurement, such as temperature.

Negative numbers are first mentioned in a Chinese text called *Jiuzhang suanshu* (The Nine Chapters on the Mathematical Art). It was compiled by several authors during the period from the second century BCE to the first century CE. The Bakshali Manuscript, an Indian text of uncertain date but no later than the seventh century, also uses negative numbers, though they are confusingly indicated by a "+" sign. The minus sign was first used to show a negative number by Johannes Widmann in 1489. It is from the Indian texts that negative numbers entered western mathematics.

some things—and most measures—can be divided into portions smaller than a unit. A loaf can be broken in half, or a person can drink a third of a bottle of wine, or a place may be a quarter of a mile away. Fractions are a useful way of expressing the size of a portion. Fractions may be expressed as one number (the numerator) divided by another (the denominator), such as ¼ (one divided into four parts). A fraction is also called a rational number as it expresses the ratio between numbers—so ¼ shows the ratio 1:4. In a decimal fraction, numbers after the decimal point indicate tenths, hundredths, thousandths, and so on according to their position.

Decimal fractions can express irrational numbers—those which are not a ratio of two whole numbers and have an infinite number of digits after the decimal point. Irrational numbers caused problems for many early mathematicians (see page 56). The Persian mathematician and poet Omar Khayyam (1048–1131) accepted all positive numbers, rational and irrational.

PARTS AND WHOLES

Some things are not divisible—we can't speak of two and a half people or three quarters of a grain of sand. But

Many things can be broken down into smaller and smaller portions—but a loaf broken into crumbs is of little use.

> *"Sexagesimals and sixties are to be used sparingly or never in mathematics, and thousandths and thousands, hundredths and hundreds, tenths and tens, and similar progressions, ascending and descending, are to be used frequently or exclusively."*
>
> François Viète, 1579

Under and Over the Bar

The Egyptians used a uniquely perplexing form of fractions around 1000 BCE, and the Greeks followed their example 500 years later. In India, Jain mathematicians wrote of operations with fractions in the *Sthananga Sutra, ca.*150 BCE.

The modern way of writing fractions with a bar or vinculum dividing the numerator and denominator stems from the Hindu method of writing one numeral above the other, used in the *Brahma-sphuta-siddhanta* (*ca.*620). Arab mathematicians added the bar to separate the two figures. The first European mathematician to use the fraction bar as it is used now was Fibonacci (*ca.*1170–1250).

Getting to the Point

Decimal fractions were recorded in Indian units of measurement around 2800 BCE. Weights found in the archeological remains of a settlement called Lothal in the Indus Valley area (modern-day Gujarat) weigh 0.05, 0.1, 0.2, 0.5, 1, 2, 5, 10, 20, 50, 100, 200, and 500 units (a unit being around 28 grams or one ounce). Abu'l Hasan Ahmad ibn Ibrahim al-Uqlidisi (*ca.*920–80) wrote the first known Arab text on the use of decimal fractions. In the 12th century, the Iranian mathematician Ibn Yahya al-Maghribi al-Samaw'al provided a systematic treatment of how to use decimal fractions to give approximate values for irrational numbers.

Decimals came quite late to Europe. Francesco Pellos wrote a treatise published in Italy in 1492 that seems to use a decimal point to divide units from tenths, but his work does not show a rigorous understanding of what he had done. Christoff Rudolff, writing in a German accountancy text in 1530, was the first to show a thorough understanding of how to work with decimal fractions, though he used

The statue of Simon Stevin in Bruges. Apart from his work in science and mathematics, he invented the first land yacht, which could travel as fast as a horse.

a vertical bar instead of a decimal point. The first European treatise on decimals was produced by Simon Stevin in 1585, and he is generally credited with introducing decimal fractions into Europe. Stevin used a different notation from that we use now, writing 5.912 in the form:

5 ⓪ 9 ① 1 ② 2 ③

The French mathematician François Viète (1540–1603) experimented with several ways of writing decimal fractions. He tried raising and underlining the fractional part (627,125 $^{\underline{512,44}}$), and showing the decimal as a fraction (627,125 $\frac{512,44}{1,000,00}$), using a vertical stroke to separate the integral and decimal parts (627,125 | 512,44), and showing the integral part in bold type (**627,125**,512,44). But he was not to come up with the method that has stuck. The earliest printed use of the decimal point was by Giovanni Magini (1555–1617), an Italian map-maker,

François Viète worked for the court of Henry of Navarre. By cracking a cipher used by the Spanish, he enabled the French to read enemy despatches.

EGYPTIAN FRACTIONS

The Egyptians had a strange way of working with fractions. They had special characters for half, , and two-thirds, . Thereafter, a fraction was shown by the character  written above the denominator, which was shown using the usual Egyptian symbols for numbers. So  means $^1/_7$.

However, with the exception of ⅔, the Egyptians only used unitary fractions (those with a numerator of 1); there was no way to show a numerator, so it was impossible to write ⅖ or $^3/_7$. To complicate matters further, it was not allowed to repeat a fraction—so ⅖ could not be written as ⅕ + ⅕. Instead, it was necessary to find a way of making ⅖ from unique fractions:

$$^2/_5 = {^6/_{15}} = {^5/_{15}} + {^1/_{15}} = {^1/_3} + {^1/_{15}}$$

GOOGOL AND GOOGOLPLEX

The terms "googol" and "googolplex" were invented by Milton Sirotta (1911–1981), the nine-year-old nephew of American mathematician Edward Kasner (1878–1955). A googol is 1 followed by 100 zeroes; a googolplex is 1 followed by a googol zeroes.

These numbers are inconceivably large. There are fewer than a googol fundamental particles in the known universe (fundamental particles are subatomic; there may be 10^{81} in the universe). If you could write down a googolplex in standard 10 point type, it would be 5×10^{68} times longer than the diameter of the known universe, and writing at two digits per second would take 10^{82} times the age of the universe to complete.

17 being zeroes and the last 1. This can be extended to show other numbers as multiples of a power of ten. For instance, $10^3 = 1,000$ and so $6.93 \times 10^3 = 6,930$ and $6.93 \times 10^{-3} = 0.00693$. Scientific notation is much easier to understand, and more compact to write, than a long string of digits.

The first use of scientific notation is not known, but it was already current in 1863 when an encyclopedia included the following text:

astronomer, and friend of Kepler, who used it in 1592. Even so, it did not catch on until John Napier used it in his tables of logarithms over twenty years later. Napier suggested in 1617 that the full stop or comma could be used, and settled on the full stop in 1619, though many European countries have adopted the comma as their decimal separator.

BIGGER AND BIGGER

While fractions and decimals provide a way of writing very small numbers, developments in science have led to a need for ways of representing and talking about increasingly large numbers.

Scientific notation uses powers of ten to show both very large and very small numbers. A power of ten shows how many figures come before or after the decimal point. For example, 10^{18} is 1 followed by 18 zeroes. In the other direction, 10^{-18} is a decimal point followed by 18 digits, the first

"a current force equal to 10,000,000,000 times the value given by the quotient of 1 meter by 1 second of time, that is, 10^{10} meter/seconds."

John Napier, the inventor of logarithms, believed that the world would come to an end in either 1688 or 1700.

THE SAND RECKONER

In what was effectively one of the world's first research papers, Archimedes boasted in the third century BCE that he could write a number larger than the number of grains of sand it would take to fill the universe. He was able to do this using the new Ionian number system and his own notation, which in effect used powers and was based on the "myriad," or 10,000. He worked with powers of a myriad myriad, or 100,000,000. Archimedes' estimate of the size of the universe, while far larger than previous figures, was nowhere near modern estimates. His number of grains of sand was 8×10^{63}.

SCIENTIFIC NOTATION	US NAME	EUROPEAN NAME
10^3	Thousand	Thousand
10^6	Million	Million
10^9	Billion	1000 million (billion)
10^{12}	Trillion	Billion
10^{15}	Quadrillion	1000 billion
10^{18}	Quintillion	Trillion
10^{21}	Sextillion	1000 trillion
10^{100}	Googol	Googol
10^{303}	Centillion	–
10^{600}	–	Centillion
10^{googol}	Googolplex	Googolplex

10 to power 10^{100}, or 1 followed by googol zeroes

The use of scientific notation also gets around the confusion over the different names used for large numbers in the United States and elsewhere. Although names are the same up to a million, they then diverge. The American billion is only a thousand million (10^9), while a European billion is a million million (10^{12}) and 10^9 is just called a thousand million. The pattern continues with even larger numbers.

As science and mathematical proofs demand writing ever larger numbers, even scientific notation becomes unwieldy and finally unmanageable. Solutions to the problem include using ^ or → to indicate powers of powers, and even using polygonal shapes to indicate powers.

In 1976 Donald Knuth proposed a notation using ^ to indicate powers. The expression n^m means "raise n to the power of m (n × n m times)"

n^2 = n^2	3^2 is 3^2 = 3 x 3 = 9
n^3 = n^3	3^3 is 3^3 = 3 x 3 x 3 = 27
n^4 = n^4	3^4 is 3^4 = 3 x 3 x 3 x 3 = 81

Doubling the ^ symbol to n^^m means "calculate n^n m times" and n^^^m means "n^^m m times."

So while

3^3 is 3^3 = 27

3^^3 is 3^(3^3) = 3^{27} = 7,625,597,484,987

And tripling the ^ symbol to ^^^ rapidly leads to very large numbers:

3^^^3 is 3^^(3^^3) =
$3^{7,625,597,484,987}$^$3^{7,625,597,484,987}$

THE LARGEST NUMBER EVER

The largest number that has been cited in any theoretical mathematical problem is called Graham's Number, named after American mathematician Ronald Graham (b. 1932). It was devised by Graham as the upper bound of a possible solution to a problem. The number is so large that it is impossible to write it in any of the notational forms covered here. It is said that if all the matter in the universe were turned into ink it would not be enough to write the number out in full. Ironically, experts suspect that the real answer to the original problem is "6."

As the number of ^ characters increases, the numbers get harder to read (as well as unimaginably large). John Conway (*b*.1937) suggests condensing the numbers by using right arrows →to indicate the number of ^ characters. So

n^^^4 would be written n→4→3.

Another way, called tetration, expresses

$$n^{n^{n^{n}}} \text{ as } {}^{4}n.$$

So $^{4}2$ is $2^{2^{2^{2}}} = 2^{2^{4}} = 2^{16} = 65{,}536$

Another system, Steinhaus-Moser notation, uses polygonal shapes to show how many times a number must be raised to a power.

△n (a number *n* in a triangle) means n^{n}.

So △2 = 2^{2} = 4, △3 = 3^{3} = 27

□n (a number *n* in a square) is equivalent to "the number n inside n triangles, which are all nested." At each stage, the number is evaluated and used for the next stage, so 2 in a square is 2 in two nested triangles. The first nested triangle is 2^{2} = 4, so the next nested triangle is 4^{4} = 256.

⬠n (a number *n* in a pentagon) is equivalent to "the number n inside n squares, which are all nested." Originally, this was the limit of Steinhaus's system and he used a circle for this: Ⓝ.

② starts from 256^{256} and evaluates this in the same way 256 times. Steinhaus gave the number ② the name a *mega*, and ⑩ the name *magiston*. Moser's number is 2 inside a polygon with mega sides.

MOVING ON

Now that we are equipped with large enough numbers, we can begin to put them to work. What numbers can do on their own is the subject of pure mathematics; what they can do when they are recruited into the service of other disciplines is applied mathematics. A culture must develop at least a little pure mathematics before it can start applying numbers to real-world problems such as building, economics, and astronomy, so we will start with number theory.

CHAPTER 2

NUMBERS
Put to Work

Counting is a good start, but any more sophisticated application of numbers requires calculations. The basics of arithmetic—addition, subtraction, multiplication, and division—came into use early on through practical applications.

As soon as people started to work with numbers in this way, they began to notice patterns emerging. Numbers seem to play tricks, to have a life of their own, and to be able to surprise us with their strange properties. Some are simple but elegant—like the way we can multiply a two-digit number by 11 simply by adding the digits together and putting the result in the middle: $63 \times 11 = 693$ (6 + 3 = 9, put 9 between 6 and 3).

Some are breathtaking in their sophistication. Number theory, which includes arithmetic, is concerned with the properties of numbers. Ancient people imbued numbers with special powers, making them the center of mystical beliefs and magical rituals. Modern mathematicians talk of the beauty of numbers.

A man uses an abacus in a Japanese sword shop, ca.1890.

Putting Two and Two Together

The rules of arithmetic provided the ancients with methods for working out fairly simple sums, but as the numbers involved grew larger, tools to help with—and eventually to mechanize—calculation become increasingly important. Tools to simplify addition, subtraction, multiplication, and division emerged very early on. Over the last few centuries these simple aids have not been sufficient and our tools for working with numbers have become increasingly complex and technically sophisticated, until we now have computers that carry out in a fraction of a second calculations that would have seemed quite inconceivable to the earliest mathematicians.

A Soviet schoolboy uses a Russian abacus—the schoty—during a math lesson in 1920. Abacuses are still widely used in Russia, but no longer in schools.

STRINGS, SHELLS, AND STICKS

The earliest mathematical tools were counting aids such as tallies and beads, shells, or stones. The Yoruba in west Africa used cowrie shells to represent objects, always reckoning them in groups of 5, 20, or 200, for example. Other civilizations have used different objects.

In Meso-America, the Inca civilization had no written number system but used *khipu* (or *quipu*)—groups of knotted strings—to record numbers. A *khipu* consists of colored strands of alpaca or llama wool, or sometimes cotton, hanging from a cord or rope. It could be used to record ownership of goods, to calculate and record taxes and census data, and to store dates. The strings could be read by Inca accountants called *quipucamayocs,* or "keepers of the knots." Different-colored strands were apparently used to record differing types of information, such as details relating to war, taxes, land, and so on.

Shells have been used as counting aids and as currency.

KNOTTY PROBLEMS

The position of a group of knots on a *khipu* shows whether that group represents units, tens, hundreds, etc. Zero is indicated by a lack of knots in a particular position. Tens and powers of ten are represented by simple knots in clusters, so 30 would be shown by three simple knots in the "tens" position. Units are represented by a long knot with a number of turns that represents the number, so a knot with seven turns shows a seven. It's impossible to tie a long knot with one turn, so one is represented by a figure-of-eight knot. *Khipus* recorded information such as population censuses or details of crops harvested and stored.

Although it looks like a decorative fringe, the khipu *was a sophisticated accounting aid. This one was made in Peru ca.1430–1532.*

North American tribes also used knotted strings, called *wampam*, and knots in leather straps have been used in less sophisticated arrangements by the Persians, Romans, Indians, Arabs, and Chinese.

In Papua New Guinea, tally ropes were used to record the trade in gold lip pearl shells. In Germany, bakers used knotted ropes to tally bakery orders until the late 19th century. Herdsmen in Peru, Bolivia, and Ecuador used a form of *khipu*, with groups of white strings for sheep and goats and green strings for cattle, until the 19th century.

The practice has proved remarkably enduring. In Tibet, knotted prayer strings still help Buddhists to keep track of their prayers; the same function is performed by Muslim prayer beads and Catholic rosaries.

TIMES TABLES

Tables of numbers for looking up the results of calculations, particularly multiplication, have been used for thousands of years. Clay tablets dating from around 1800 BCE preserve ancient multiplication tables used in Mesopotamia. The idea of compiling tables of the results of common arithmetic operations is as old as written mathematics. The mathematicians of ancient Babylon inscribed their work on clay tablets; many of these present mathematical tables for multiplication, squares and cubes and their roots, and reciprocals.

A clay plate from ca. 2500 BCE, found at Lagash in Iraq, contains a record of numbers of goats and sheep.

BEADS AND BOARDS

Some cultures developed quite ingenious tools and systems to help with calculations. One of the more familiar is the abacus, developed around 3000 BCE in Mesopotamia and still in use in some eastern cultures. It began life as a board or slab covered in sand, used in ancient Babylon for aligning numbers or writing; it later developed into a board with lines or grooves for counters. The modern abacus with counters threaded on to rods or wires requires more technological advancement to produce, but is used in much the same

SHOW ME THE MONEY

The name Chancellor of the Exchequer for the minister in charge of the country's finance in the UK comes from the use of the "exchequer" board—a counting board similar in design to a chess board used as an abacus.

The Exchequer was the medieval English institution charged with collecting royal revenues.

LOGARITHMS

Logarithms offer a quick way of carrying out long division and multiplication. They work on the principle that to multiply powers we can add them together.

$10^1 = 10$
$10^2 = 100$
$10^1 \times 10^2 = 1,000 = 10^3$
Looking at the powers: $1 + 2 = 3$

The logarithm of a number n is the power to which the base number (in this case, 10) is raised to give n. So the logarithm of 10 is 1 because $10^1 = 10$; the logarithm of 100 is 2, because $10^2 = 100$. The logarithm of 2 is 0.30103 because $10^{0.30103} = 2$.

Any two numbers can be multiplied together by adding their logarithms. So $\log_{10}10 + \log_{10}100 = \log_{10} 1,000$. The subscript shows that we are using logarithms to base 10—i.e., working with powers of 10.

The same principle of working with powers obviously holds with other numbers besides 10:

$2^4 \times 2^{10} = 2^{14}$
$(16 \times 1,024 = 16,384)$

So using logarithms with a base of 2, the logarithm of 16 would be 4. Logarithms can be constructed to any base.

century and is still used there as well as in the Middle East and China. Earlier Chinese mathematicians used rods of different lengths which they laid out in matrix on a special table or board. The principle was similar to the abacus in that the position of the rods indicated their value. In Europe, merchants continued to use the abacus until at least the 17th century, when it was replaced by arithmetic algorithms following the ascendance of Hindu-Arabic numerals.

Certain early Arab mathematicians took over the basic algorithms for calculation from India, and around 950 Abu'l Hasan Ahmad ibn Ibrahim al-Uqlidisi adapted them for use with pen and paper rather than the traditional Indian dustboard.

TRICKIER CALCULATIONS

As both science and commerce became more advanced and sophisticated, the need to work with large numbers, fractions, and decimals increased. Calculations became hard and time-consuming and people searched for ways to make them more manageable. The most ingenious and enduring solution was the development of logarithms by the Scottish mathematician John Napier in the early 17th century.

way. The position of a bead or counter denotes whether it stands for a unit, a ten, a hundred, and so on. A practiced, proficient user can move the beads or counters at great speed, carrying out calculations as quickly as many later mechanical calculators. As late as the 1920s, accountants training in the City of London had to be able to use an abacus as well as arithmetical methods.

The abacus spread to Japan in the 16th

JOHN NAPIER (1550–1617)

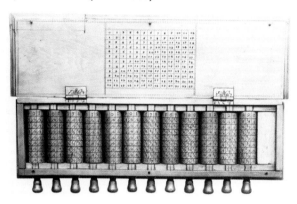

John Napier was a Scottish mathematician and eighth Laird of Merchiston. He entered the University of St Andrews at the age of 13, but left without a degree. He is best known as the inventor of logarithms and another calculating device called "Napier's bones." He began working on logarithms around 1594 and published his treatise, *Description of the Marvelous Canon of Logarithms,* in 1614. Napier's bones comprised a system of small rods used for calculating; they were the forerunner of the slide rule.

Napier was also an inventor of artillery, and suggested to James VI of Scotland something like a tank—a metal chariot with holes from which small bore shot could be fired. He is known, too, as the first person

The rods in Napier's Bones carried multiplication tables which made calculations much simpler, but they did not work in the same way as logarithms.

to use the dot as a decimal point separating the parts of a decimal number—his logarithmic tables are the first document to use the decimal point in the modern style. He was ardently anti-Catholic and believed the Pope to be the Antichrist.

Tables of logarithms were published first in 1620 by the Swiss mathematician Joost Bürgi, who discovered logarithms independently of Napier between 1603 and 1611. To use logs it was necessary first to look up the logarithms of the numbers to be multiplied, then add them together, and finally to look up the antilogarithm of the answer. For division, it was necessary to subtract one logarithm from the other and then look up the antilog.

Logarithms also offer an easy way of finding powers and roots. To find a square, multiply the logarithm by two and look up the antilog; to find a square root, divide the log by two and look up the antilog. To find a cube, the logarithm is tripled; to find a cube root, it is divided by three, and so on. Children in western schools were taught to use tables of logarithms until the late 20th century, when electronic calculators finally took over the role of complex calculations.

The development of logarithms made much else possible. For scientists, the complex calculations required, particularly for astronomy, became much easier and so progress in this field speeded up. It didn't take long before logarithms moved from printed tables to physical calculating devices. The first was the Gunter scale, developed by Englishman Edmund Gunter in 1620. It was a large plane scale with logarithms printed on it. Alongside a pair of compasses, sailors used it to multiply and divide distances.

Although base-10 logarithms are barely used any more (their function has been taken over by calculators and computers), base-e logarithms (natural logarithms; see panel) are still widely used in science.

An enduring mechanical calculator that used logarithms was the slide rule. The first slide rule was circular and designed by William Oughtred around 1632; he made a rectangular version in 1633. The slide rule has decimal numbers on one scale and their logarithms on another. By lining up the scales in the right way, it's possible to read off the product of two numbers.

MACHINES FOR MATHEMATICS

Charts and tables, and then the slide rule, offered a great advantage over carrying out calculations with paper and pen, but the enormous burden of calculation required by emerging science, especially astronomy, by commerce, finance, and navigation cried out for better mechanical aids.

The first commercial attempt at a calculating machine was made by Blaise Pascal (see page 43) in 1642–3 to help his father, an administrator in Rouen, France,

---

ALL ABOUT e

e is a very significant number in mathematics. It is defined (among other methods) as the sum of all numbers in the series

$$\frac{1}{0!} + \frac{1}{1!} + \frac{1}{2!} + \frac{1}{3!} + \frac{1}{4!} + \dots$$

where n! means n-factorial (n multiplied by each digit smaller than itself—so

$$4! = 4 \times 3 \times 2 \times 1 = 24).$$

By convention, 0! = 1. So

$$e = 1 + 1 + {}^{1}/_{2} + {}^{1}/_{6} + {}^{1}/_{24} + {}^{1}/_{120} + \dots$$

It is an example of an infinite series as the number of terms on the right-hand side is unending.

Starting in the late 17th century, the slide rule reigned for 300 years as the king of calculators. It was superseded by the pocket calculator in the 1970s.

who had to deal with complicated tax figures. The Pascaline, as it was called, consisted of a box containing a series of notched wheels or gears. A complete rotation of one wheel advanced the adjacent wheel one tenth of a rotation. It could only carry out addition and subtraction, and was not hugely useful as French currency at the time was not decimal: there were 12 *deniers* to a *sol* and 20 *sols* to a *livre*. (This is the same as the system in use in the UK until 1970, which had 12 pence to a shilling and 20 shillings to the pound.)

A slightly earlier machine that also used rotating cogs was the Calculating Clock designed by Wilhelm Schickard (1592–1635) in 1623. He made a prototype, which was destroyed in a fire, and possibly a second copy which has never been found. Schickard died of bubonic plague and his invention was lost to history. However, he described his invention in papers (including letters to the astronomer Johannes Kepler) which were discovered in the 20th century, enabling his machine to be recreated in 1960.

Gottfried Leibniz (see page 154) developed the principle of Pascal's machine into a fully functional calculating machine that could handle addition, subtraction, multiplication, and division. Like Pascal, Leibniz was a child prodigy. He had learned Latin by the age of 8 and gained his second doctoral degree at 19. His first "Stepped Reckoner" prototype was built in Paris in 1674. It used a central cylinder with a set of rod-shaped teeth of different lengths that extended along the drum. This turned a series of toothed wheels. Despite his genius,

> *"It is unworthy of excellent men to lose hours like slaves in the labor of calculation which would safely be relegated to anyone else if machines were used."*
>
> Gottfried Leibniz

Leibniz died in poverty. The prototype of his machine was ignored and lay hidden in an attic in the University of Göttingen, Germany, until 1879.

The 18th century saw a flurry of calculating machines based on the same principles as those of Pascal and Leibniz, but none really took off commercially since mechanical limitations meant that they were never quick and easy to use.

The first successful calculating machine was made by Charles Xavier Thomas de Colmar (1785–1870), in France. His Arithmometer worked on the same principle as Leibniz's machine and could carry out all four arithmetic operations easily. Between 1820 and 1930, 1,500 were sold and similar devices appeared from other manufacturers.

TOWARD A COMPUTER

The precursor of the modern computer is generally considered to be the Analytical Engine designed by Charles Babbage (1791–1871). At the time Babbage was working, complex calculations were carried out using tables of figures, including logarithms, compiled by people called "computers." The tables tended to have a lot of errors—Babbage's aim was to make a machine that could perform calculations without making mistakes. He began designing his first such machine, a

BLAISE PASCAL (1623–62)

The French mathematician, physicist, and philosopher Blaise Pascal laid the foundation of probability theory and invented the first digital calculator. His mother died when he was a small child and the family moved to Paris, where his father took on his son's education. Pascal was something of a prodigy, publishing his first paper on mathematics at the age of 18. As well as designing his calculating machine, he worked on pressure and hydraulics, formulating Pascal's law of pressure and making a mercury-filled barometer. In his thirties, he underwent an intense religious experience, adopted the strict moral code of Jansenism, and entered the convent of Port-Royal in 1655, giving up his interest in mathematics.

The Pascaline could deal with numbers up to 9,999,999, but could only be used for addition and subtraction. It was operated by moving the dials, the solution to the problem appearing in the windows above.

Difference Engine, in 1822 to work out the values of polynomial functions (functions that contain more than one term, such as $4x^2 + 5x$). The first Difference Engine he designed needed around 25,000 parts weighing a total of 15 tons (13,600 kilograms). It would have stood 8 feet (2.4 m) tall. He never built it, but designed an improved version, Difference Engine No. 2.

COW CATCHER

Babbage also invented the cow-catcher—the metal frame attached to the front of trains to clear the track of obstacles.

> *"The Analytical Engine weaves algebraic patterns, just as the Jacquard loom weaves flowers and leaves."*
>
> Ada Lovelace

Difference Engine No. 2 was constructed by the Science Museum in London to celebrate the 200th anniversary of Babbage's birth. It worked flawlessly.

THE JACQUARD LOOM

The mechanical Jacquard loom, invented by Joseph Marie Jacquard in France in 1801, uses punched cards to store a woven pattern and control the loom to reproduce the pattern. It was the first piece of machinery to be controlled by punched cards and, although it was entirely mechanical rather than computerized, it is considered an important step toward computer programing.

Again, Babbage did not make it, but it was built to his design in the Science Museum in London in 1989–1991, to the engineering tolerances of Babbage's time. It produced a solution accurate to 31 digits at its first trial.

Babbage abandoned plans for the Difference Engine and embarked on a more ambitious project—to design an Analytical Engine that could accept programed instructions on punched cards. Again he did not actually build it, but refined the design repeatedly. The mathematician Ada Lovelace read of his design and constructed

BERNOULLI NUMBERS										
n	0	1	2	4	6	8	10	12	14	16	18
B	1	$-\frac{1}{2}$	$\frac{1}{6}$	$-\frac{1}{30}$	$\frac{1}{42}$	$-\frac{1}{30}$	$\frac{5}{66}$	$-\frac{691}{2,730}$	$\frac{7}{6}$	$-\frac{3,617}{510}$	$\frac{43,867}{798}$

a program to calculate Bernoulli numbers using the Analytical Engine. Bernoulli numbers are a sequence of positive and negative rational numbers important in number theory and analysis (see table above).

We have come a long way from Babbage's plans for machines that filled a room and performed only arithmetic, though there is some debate about who created the very first true computer.

AUGUSTA ADA KING, COUNTESS OF LOVELACE (1815–52) —"PRINCESS OF PARALLELOGRAMS"

Augusta Ada King, often known as Ada Lovelace, was the daughter of British poet Lord Byron and Annabella Milbanke; the couple separated two months after her birth. Her mother hoped that instruction in mathematics might root out any madness Ada could have inherited from her father and engaged Augustus De Morgan, first professor of mathematics at University College, London, to teach her.

Ada first took an interest in Babbage's work around 1833. During a nine-month period in 1842–3, she translated the work of Italian mathematician Luigi Menabrea on Babbage's Analytical Engine. Her instructions for calculating Bernoulli numbers using the Engine are widely considered to be the world's first computer program, even though they were never actually implemented. Ada worked with Babbage until her premature death, which was caused by overenthusiastic bloodletting on the part of doctors trying to treat her for cancer of the uterus.

The German engineer Konrad Zuse (1910–95) made the first binary computer, the Z3, in 1941 but it was only partially programmable. The first completely programmable computer was the Colossus, designed by Tommy Flowers (1905–98) for the UK Secret Service during World War II and used to crack the high-level codes of the German army. The first Colossus went into service in early 1944 and ten had been built by the end of the war. However, most were destroyed at the end of the war and for many years the British government refused to acknowledge that they had ever been

> ### CHIPS, ANYONE?
> The microchip was first invented in 1952 by a Ministry of Defence worker in the UK, Geoffrey Dummer. However, the MOD refused to fund development and a patent was filed in the United States by Jack Kilby seven years later.

built. Stepping into the gap left by the disowned Colossus, the United States claimed the first computer with the ENIAC designed by John Mauchly and J. Presper Eckert and completed in 1946. After the war, enormous mainframe computers were produced for use in industry and by governments and other large organizations and the commercial computer industry was born. These early computers cost hundreds of thousands of dollars and worked with punched tapes or cards. They had no screen or keyboard and were used largely for scientific, military, and financial applications.

A COMPUTER IN THE HAND...

Computers for everyone became a reality with the development of the microchip in 1958 by Jack Kilby, who worked for Texas Instruments in the United States. A microchip, or integrated circuit, packs all the circuitry required for a complex electrical system on to a tiny wafer of silicon using etching technology. It enabled the miniaturization of the previously huge machines used for calculations.

The microchip led to the introduction of the first handheld calculators in 1970 and then personal computers. The Intel 4004, developed in 1971, was the first microchip to put all the functions of a computer on a

Despite filling a room, Colossus had less processing power than an iPod. After the war, the British government denied it had ever existed.

single chip, which heralded the revolution in computer design. The number of instructions that could be fitted on to a microchip doubled every year as manufacturing advanced. A modern microchip can hold features smaller than a micrometer across (one millionth of a meter, or a thousandth of a millimeter). Microchips are everywhere, controlling our planes, cars, and household appliances—they are even cheap enough to put into birthday cards that play a tune when opened.

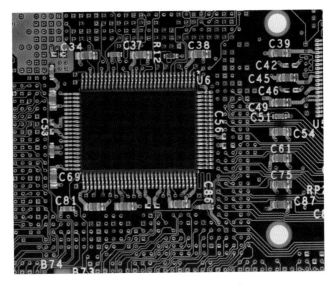

The circuitry on a microchip is too small to see with the naked eye. Microchips are everywhere, controlling nearly all our technologies.

The speed and power of computers continues to increase at an astonishing rate. The computers used to put a man on the moon in 1969 could be outwitted by a mobile phone today.

Our fastest supercomputers carry out hundreds of trillions of operations a second, which is around a million times faster than a standard desktop computer. Top500 is the name of a project that ranks the 500 fastest and most powerful non-distributed computer systems in the world.

The demands we place on computers are ever-increasing, too. Decoding DNA, analyzing radiation from outer space for telltale signs of a deliberate message, and rendering digital movies at the highest resolutions still demand hours and days of dedicated computer time. The next generation of computers may leave silicon circuitry behind altogether and move toward quantum computing, making use of the subatomic properties of matter to store and manipulate data.

Early calculating aids intended to speed up arithmetic. Computers first did this, facilitating bulk calculations, saving time, and giving accurate results. Computers can deal with extremely complicated tasks as long as the tasks can be specified in logical steps. This has provided an impetus for advances in logic and its notation, and even for computers to handle these tasks. Now, computers are used to manipulate mathematical expressions in symbolic form, working directly with algebraic equations rather than calculating with numbers fed into equations. For example, the Schoonschip program, developed by Nobel physicist Martinus Veltman (born 1931), handles the calculations required for high-energy physics.

Special Numbers and Sequences

People have long been fascinated by the apparently magical abilities of numbers to fall into patterns and to throw up surprising rules. Some of these became apparent to very early mathematicians. These numbers were often incorporated into mystical or religious rituals, buildings, and artifacts. The strange properties of numbers are now the domain of number theorists.

PRIME NUMBERS

Primes are a special class of integers: they are numbers that have no factors (cannot be divided by anything) except themselves and 1. The primes under 20 are 2, 3, 5, 7, 11, 13, 17, and 19 (1 is usually not included). As numbers get larger, primes become less frequent, but remain surprisingly common. Even with numbers around 1,000,000 about 1 in 14 is prime. People have studied prime numbers for millennia, originally ascribing some mystical or religious significance to them. Around 300 BCE the Greek mathematician Euclid was the first to prove that there is an unending sequence of primes. Still, more than 2,000 years later, we have no formula for predicting primes.

Prime numbers sound as though they are nothing special—perhaps rather defective, even, having no real factors. But there are clusters of interesting phenomena around them and they have become central to number theory.

FINDING PRIMES

It's easy to find small prime numbers; we can all do it in our heads. But finding larger primes becomes increasingly difficult.

Prime number theory attempts to predict the frequency of primes. The French mathematician Adrien-Marie Legendre (1752–1833) conjectured in 1798

Euclid presenting his work to King Ptolemy I Soter in Alexandria. The illustration is by Louis Figuier and dates from 1866.

THE SIEVE OF ERATOSTHENES

The Ancient Greek mathematician Eratosthenes (276–194 BCE) developed a simple algorithm for finding prime numbers, called the sieve of Eratosthenes.

How to sieve primes:

1. Begin by drawing up a square grid containing all the numbers from 1 to your top limit for primes. Cross out or color in 1—it's not a prime.

2. The first prime is 2; write this at the top of your list. Color multiples of 2.

3. The next remaining number is the next prime (3), so write this in the list of primes. Color multiples of 3.

4. The next remaining number is the next prime (5); write this in the list of primes and color multiples of 5.

5. Continue to the end of the square. The numbers in the list are the primes.

First grid:

1	2	3	4	5	6	7	8	9	10
11	12	13	14	15	16	17	18	19	20
21	22	23	24	25	26	27	28	29	30
31	32	33	34	35	36	37	38	39	40
41	42	43	44	45	46	47	48	49	50
51	52	53	54	55	56	57	58	59	60
61	62	63	64	65	66	67	68	69	70
71	72	73	74	75	76	77	78	79	80
81	82	83	84	85	86	87	88	89	90
91	92	93	94	95	96	97	98	99	100
101	102	103	104	105	106	107	108	109	110
111	112	113	114	115	116	117	118	119	120

PRIME NUMBER
2

Second grid:

1	2	3	4	5	6	7	8	9	10
11	12	13	14	15	16	17	18	19	20
21	22	23	24	25	26	27	28	29	30
31	32	33	34	35	36	37	38	39	40
41	42	43	44	45	46	47	48	49	50
51	52	53	54	55	56	57	58	59	60
61	62	63	64	65	66	67	68	69	70
71	72	73	74	75	76	77	78	79	80
81	82	83	84	85	86	87	88	89	90
91	92	93	94	95	96	97	98	99	100
101	102	103	104	105	106	107	108	109	110
111	112	113	114	115	116	117	118	119	120

PRIME NUMBER
2 3 5

Third grid:

1	2	3	4	5	6	7	8	9	10
11	12	13	14	15	16	17	18	19	20
21	22	23	24	25	26	27	28	29	30
31	32	33	34	35	36	37	38	39	40
41	42	43	44	45	46	47	48	49	50
51	52	53	54	55	56	57	58	59	60
61	62	63	64	65	66	67	68	69	70
71	72	73	74	75	76	77	78	79	80
81	82	83	84	85	86	87	88	89	90
91	92	93	94	95	96	97	98	99	100
101	102	103	104	105	106	107	108	109	110
111	112	113	114	115	116	117	118	119	120

PRIME NUMBER
2   3   5   7
11 13 17 19
23 29 31 37
41 43 47 53
59 61 67 71
73 79 83 89
97 101 103 107
109 113

that the number of primes below a number x is approximately

$$x/\ln(x) - 1.08366$$

where $\ln(x)$ is the natural logarithm of x. As the size of x increases, the error involved becomes comparatively smaller.

The chances of a number, x, being a prime are

$$1/\ln(x).$$

This means, for instance, that a number

ERATOSTHENES (276–194BCE)

Eratosthenes was born in Libya, but worked and died in Alexandria (Egypt). A friend of Archimedes, he was in charge of the library at Alexandria. Around 250 BCE he invented the armillary sphere, a spherical model with intersecting bands that is used to demonstrate and predict the movement of the stars. It was used as an astronomical instrument until the 18th century.

Eratosthenes also developed a system for measuring longitude and latitude, drew a map of the whole known world, and made the first recorded calculation of the Earth's circumference (see panel, page 87: Measuring the Earth).

The later writer Eusebius of Caesarea (d.339–40 CE) attributes to Eratosthenes a calculation of the distance from the Earth to the sun which is accurate to within 1 percent of the figure now accepted.

The Earth is represented by the ball at the center of the armillary sphere, the apparent orbits of other bodies by the rings around it.

around 1,000,000 has a chance of about 1 in 13.8 of being prime since the natural logarithm of 1,000,000 is 13.8.

TWIN PRIMES

Twin primes are pairs of prime numbers separated by only 2. Obvious examples are 3 and 5, 5 and 7, 11 and 13, or 17 and 19. The twin primes conjecture states that there is an infinite number of twin primes. That seems reasonable, as it only means they don't have to run out at some point. But it hasn't been proven to be true. There is also a "weak" twin primes conjecture, which has been demonstrated. This states that the number of twin primes below a number x is approximately given by this horribly complicated expression:

THE GOLDBACH CONJECTURE

In 1742, the Prussian mathematician Christian Goldbach wrote a letter to the Swiss mathematician and physicist Leonhard Euler in which he set out his belief that every integer greater than 2 can be written as the sum of three primes. He considered 1 to be a prime, which mathematicians no longer accept. The conjecture has since been refined and now states that every even number greater than 2 can be written as the sum of two primes.

Goldbach could not prove his belief (which is why it is a conjecture and not a theorem), and no one has been able to prove it since. It has been verified by computer for all numbers up to 10^{18}, but a theoretical proof is still needed.

There wasn't a great deal of small talk in the letter Goldbach wrote to Euler, but that's how it is with mathematicians.

$$2\,\pi \sum_{p\geq 3} \frac{p(p-2)}{(p-1)^2} \int_2^x \frac{\mathrm{d}x}{(\log x)^2} = 1.320323632 \int_2^x \frac{\mathrm{d}x}{(\log x)^2}$$

Don't worry about the expression—it doesn't matter. What is interesting to think about is why it exists at all. What is it about numbers that makes it possible to find an expression like this? The number in the middle, 1.320323632, is called the prime constant. It has no other known relevance except in this prediction of twin primes.

PERFECT NUMBERS

Perfect numbers are those that are the sum of all their proper divisors. This means that if you add together all the numbers that the number can be divided by, the answer is the number itself.

For example
6 = 1 + 2 + 3
28 = 1 + 2 + 4 + 7 + 14

Euclid first proved that the formula $2^{n-1}(2^n-1)$ gives an even perfect number whenever

MIHĂILESCU'S THEOREM

In 1844, Belgian mathematician Eugène Charles Catalan (1814–94) conjectured that $2^3 = 8$ and $3^2 = 9$ form the only example of consecutive powers (i.e., 2 and 3, with cube and square, 8 and 9). It was finally proven to be the case by the Romanian mathematician Preda Mihăilescu in 2002.

$2^{n}-1$ is prime. There are currently 46 perfect numbers known; the largest is $2^{43,112,608} \times (2^{43,112,609} - 1)$ with 25,956,377 digits.

AMICABLE NUMBERS

Amicable numbers come in pairs. The proper divisors of one of the pair, added together, produce the other. The numbers 220 and 284 are an amicable pair. The proper divisors of 220 are 1, 2, 4, 5, 10, 11, 20, 22, 44, 55, and 110, which added together make 284; and the proper divisors of 284 are 1, 2, 4, 71, and 142, which together make 220. Pythagoras' followers studied amicable numbers from around 500 BCE, believing them to have many mystical properties.

Thabit ibn Qurrah (836–901), a translator of Greek mathematical texts, discovered a rule for finding amicable numbers. Arab mathematicians continued to study them, Kamal al-Din Abu'l-Hasan Muhammad al-Farisi (ca.1260–1320) discovering the pair 17,926 and 18,416 and Muhammad Baqir Yazdi finding 9,363,584 and 9,437,056 in the 17th century.

POLYGONAL NUMBERS

Some numbers of dots, stones, seeds, or other objects can be arranged into regular polygons. For example, six stones can be arranged into a perfectly regular triangle.

> "Six is a number perfect in itself, and not because God created all things in six days; rather, the converse is true. God created all things in six days because the number is perfect."
> St Augustine (354–430 CE), The City of God

Six is therefore a triangular number. If we add an extra row of stones at the bottom, we get the next triangular number, ten:

Nine stones can be arranged into a square:

The next square number has four on each side, giving $4^{2} = 16$.

Some numbers, such as 36, are both triangular and square:

Polygonal numbers are increased by incrementing each side by one extra unit.

TRIANGULAR NUMBERS

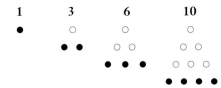

SQUARE NUMBERS

1   4   9   16

Polygonal numbers have been studied since the time of Pythagoras and were often used as the basis of arrangements for talismans. Notice how the previous triangular or square number is incremented to form the next in the series.

TRIANGULAR NUMBERS	SQUARE NUMBERS
1	1
3 (= 1+2)	4 (= 1+3)
6 (= 3+3)	9 (= 4+5)
10 (= 6+4)	16 (= 9+7)
15 (= 10+5)	25 (= 16+9)
21 (= 15+6)	36 (= 25+11)
28 (= 21+7)	49 (= 36+13)

MAGIC SQUARES

A magic square is an arrangement of numbers in a square grid so that each horizontal, vertical, and diagonal line of numbers adds up to the same total, called the magic constant. The smallest magic square (apart from a box with the figure 1

Josep Subirachs incorporated a magic square in the cathedral of the Sagrada Família in Barcelona. The magic number is 33, the supposed age of Christ at his death.

in it) has three squares on each side and the magic constant is 15:

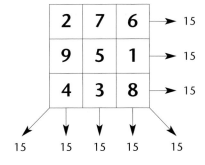

This is known as the Lo Shu square after a Chinese legend recorded as early as 650 BCE. This tells how villagers tried to appease the spirit of the flooding Lo river and a turtle came out of the water with markings on its back that depicted the magic square. The pattern acquired ritualistic or talismanic properties for the local people.

Magic squares have been known for around 4,000 years. They are recorded in ancient Egypt and India and have been attributed with special powers by cultures around the world. The first known magic squares with five and six numbers on each side are described in an Arab text, the *Rasa'il Ihkwan al-Safa* (Encyclopedia of the Brethren of Purity), written in Baghdad around 983. The first European to write about magic squares was the Greek Byzantine scholar Manuel Moschopoulos, in 1300.

The Italian mathematician Luca Pacioli, who recorded the system of double-entry book-keeping in 1494, collected and studied magic squares. (He also compiled a treatise on number puzzles and magic that lay undiscovered in the archives of the University of Bologna until it was published in 2008.)

Pi

As well as numbers that form series or patterns, there are several strange and significant single numbers. The first to be discovered was pi, π. This defines the ratio of a circle's diameter to its circumference, so that the circumference is

$$\pi d$$

where d is the diameter. The value of π is a decimal number with an infinite number of digits after the decimal point. It begins 3.14159 (which is a good enough approximation for most purposes).

That the ratio of the diameter of a circle to its circumference is always the same has been known for so long that its origins can't be traced. The Egyptian Ahmes Papyrus,

*ca.*1650 BCE, uses a value of $4 \times (8/9)^2 = 3.16$ for π. In the Bible, measurements relating to the building and equipping of the temple of Solomon, *ca.*950 BCE, use a value of 3 for π.

The first theoretical calculation seems to have been carried out by Archimedes of Syracuse (287–212 BCE). He obtained the approximation

$$^{223}/_{71} < \pi < ^{22}/_{7}$$

He knew that he did not have an accurate value, but the average of his two bounds is 3.1418, an error of about 0.0002.

Later mathematicians have refined the approximation by discovering more decimal places.

e

Another strange and very significant number is e. The value of e was first discovered by Jakob Bernoulli, who tried to discover the value of the expression

$$\lim_{n \to \infty} \left(1 + \frac{1}{n}\right)^n$$

while working on calculating compound interest. When evaluated, the expression gives the series that defines e.

The first known use of the constant, represented by the letter b, is in letters from Gottfried Leibniz to Christiaan Huygens written in 1690 and 1691. Leonhard Euler was the first to use the letter e for it in 1727, and the first published use of e was in 1736. He possibly chose e as it is the first letter of the word "exponential."

e has an infinite number of digits after the decimal place, as it is defined (among other methods) as the sum of all numbers in an infinite series—see panel, page 41.

UNREAL!

The imaginary number, *i*, is defined as the square root of minus 1.

The term imaginary number was used by the French philosopher and mathematician René Descartes (1596–1650) as a derogatory term, but now means a number that involves the imaginary square root of -1:

$$i^2 = -1$$

(A negative number can't "really" have a square root as when a number is squared, whether it is positive or negative to start with, it always gives a positive result.)

A complex number z is defined as

$$z = x + iy$$

where x and y are ordinary numbers.

Gerolamo Cardano first encountered imaginary and complex numbers in the 16th century while working on Niccolo Tartaglia's solution to cubic equations, but did not consider them useful or valid. They were first described and investigated by Rafael Bombelli in 1572.

However, even negative numbers were distrusted at the time, so people had little time for imaginary numbers. It was in the 18th century that they began to be taken more seriously. It was brought to the attention of mathematicians properly by Carl Friedrich Gauss in 1832. Strangely, the special numbers come together in the expression which has been called the most startling in the whole of mathematics:

$$e^{\pi i} + 1 = 0.$$

This, known as Euler's identity, is a special case of a rule which relates complex numbers and trigonometric functions.

The Greek mathematician Pythagoras demonstrates his theorem on right-angled triangles by drawing on the ground.

Unspeakable Numbers

The concept of banning a number may seem bizarre, but it has happened for millennia and still happens even today. Some numbers have been considered just too difficult or dangerous to countenance and have been outlawed by rulers or mathematicians. But a banned number doesn't go away, it just goes underground for a while.

PYTHAGORAS' NUMBER PURGE

The ancient Greek mathematician Pythagoras did not recognize irrational numbers and banned consideration of negative numbers in his School. (An irrational number is one that cannot be expressed as a ratio of whole numbers; so 0.75 is a rational number as it is $^3/_4$ but π is irrational.) Pythagoras had to acknowledge that his ban caused problems. His theorem, which finds the length of a side in a right-angled triangle from the lengths of the other two sides, instantly runs into problems if only rational numbers are recognized. The length of the hypotenuse (longest side) of a right-angled triangle with two sides one unit long is the square root of two—an irrational number (≈ 1.414).

Pythagoras was unable to prove by logic that irrational numbers did not exist, but when Hippasus of Metapontum (born *ca*.500 BCE) demonstrated that the square root of 2 is irrational and argued for their existence, it is said that Pythagoras had him

> "It is rightly disputed whether irrational numbers are true numbers or false. Because in studying geometrical figures, where rational numbers desert us, irrationals take their place, and show precisely what rational numbers are unable to show... we are moved and compelled to admit that they are correct..."
>
> Michael Stifel, German mathematician
> (1487–1567)

drowned. According to legend, Hippasus demonstrated his discovery on board ship, which turned out to have been unwise and the Pythagoreans threw him overboard.

Pythagoras' ban was based on his esthetic and philosophical objection to the existence of irrational numbers. Later censors have had political, economic, and social reasons for trying to outlaw certain numbers or categories of number.

ARABS V. ROMANS

There was considerable resistance to the introduction of Hindu-Arabic numerals in Europe in the Middle Ages. The ease with which arithmetic could be carried out with the new number system made it attractive. As Hindu-Arabic numbers threatened to democratize numeracy they were demonized by those who had an interest in restricting numeracy and retaining it as a special tool of the elite. If mathematics were opened up to everyone, a source of power would be lost. The Catholic Church wanted to keep control of education by maintaining its hold on numbers, and in addition opposed the system from the Islamic world on religious grounds. Mathematicians who practiced the arcane systems of mathematics using an abacus were protected by the Church. So strong was the opposition to the popularization of Hindu-Arab numerals that, it is said, some poor souls were even burned at the stake as heretics for using them. However, merchants and accountants wanted to use the new system as it made their tasks easier. The battle between the algorists—those who used algorithms, or calculating methods, with Hindu-Arabic numerals—and the abacists—who used an abacus and Roman numerals—raged for centuries. The emergence of printing in Europe eventually contributed to the dominance of the Hindu-Arabic system. Dissemination of arithmetical methods became easy and it was no longer possible to contain the flow of numbers.

Eventually, of course, the establishment buckled under pressure and the number system we use now triumphed. But Roman numerals and the abacus continued to be used in some areas of life for many years.

The French were the first to release themselves from the tyranny of the abacus. After the French Revolution (1789), there was a complete reversal and use of the abacus was, in turn, banned in schools and government offices.

666—The Number of the Beast

Many religions depend heavily on number symbolism and use special numerical methods for discovering or concealing secrets. In the early years of Christianity, the Romans were using as a talisman the Magic Square of the Sun. In this magic square of six by six numbers, the numbers 1

Lady Arithmetic rejects the abacus for Hindu-Arab numerals in the Margarita Philosophica *by Gregorius Reisch, 1503.*

THE SECRET OF ZERO

The names for "zero" in use at the time when Hindu-Arabic numbers were banned in Europe were *cifra, chifre, tziphra,* and so on. These names came to stand for the whole number system that included zero. As the system was used secretly, the name also came to mean a code or secret and developed into the word "cipher."

to 36 are arranged so that the rows, columns, and diagonals add up to 111. The sum of all numbers from 1 to 36 is 666. The Church banned the possession of the magic square because in Christianity 666 is the

Number of the Beast, thought to be the enemy of God identified in the Book of Revelation. Possession of the magic square became punishable by death.

NO TO NEGATIVES

In Renaissance Europe, negative numbers were not recognized. Solutions to mathematical problems that included

Many people consider Friday 13th to be the unluckiest day of the year. The number 13 still has such negative connotations in Western culture that many tall buildings do not have a 13th floor.

negative numbers were often disregarded. Even though early Chinese and Indian mathematicians had explained the use of negative numbers, usually by relating them to economic debt, later mathematicians in Europe struggled with them. Michael Stifel (see panel, page 56) called numbers less than zero "absurd numbers," for example. The French mathematician Albert Girard (1595–1632) was probably the first major academic fully to accept negative numbers in solutions, but it took until the early 19th century for a proper foundation for arithmetic with negative numbers to be set out.

DANGEROUS DIGITS

666 is not the only specific number to have been demonized. In the United States, there is a hexadecimal (32-digit) number which has acquired the status of "illegal number." It is the key to encrypting high-definition DVDs, and its publication is technically illegal (since by using the key with the appropriate mechanism it would be possible to unencrypt the DVDs). The AACS (Advanced Access Content System) claims that it is a copyright circumvention device—possession of a copyright ircumvention device is in breach of the Digital Millennium Copyright Act (U.S., 1998). However, within a short time of its being revealed, the "secret" number was published on 300,000 web sites; attempts to remove it from the public domain were clearly going to be fruitless.

The AACS also claims to own many other numbers used for encryption but won't say what they are (as their usefulness depends on them being secret). The only "special" feature of these numbers is that they are not at all special, but are generated as random numbers. There is, understandably, considerable resistance to the notion that anyone can "own" a number and prevent others knowing or using it. Computer enthusiasts rushed to lay claim to their own numbers that they could tell everyone else not to use in retaliation and mockery of the AACS. So we can stop now. No more numbers can be used in this book as someone else owns them all!

MOVING ON

Examining numbers and their properties is all very interesting, and was good enough for the Ancient Greeks (who disdained applications of mathematics), but for most people the value of mathematics lies largely in its usefulness. Numbers allow us to measure, count, make things, run economies, and examine the universe around us. In truth, they are the key to all of science and a lot of art and have played a key role in every civilization.

Numbers rule and define the world's economies, and so impact on all our activities.

CHAPTER 3

THE SHAPE
of Things

Not everything can be counted. A herd of cattle contains a specific number of animals; even a field contains a number of blades of grass that could, theoretically, be counted. But some things can only be measured—we can't count how much water is in a pond or the distance between the hill and the sea, yet it is useful to be able to quantify them.

Geometry—working with distances, areas, and volumes in the real world—was one of the earliest applications of mathematics. (The word is derived from *geo*, "earth" and *metron*, "measure" in Greek.)

It is likely that some of the first calculations ever carried out related to building monuments, marking out land, or making artifacts for religious purposes. A first necessary step was to develop units of measure, in itself a major conceptual leap from counting. Measuring makes an artificial distinction, dividing something continuous into nominal units.

The Gold Rush produced gold to be weighed, not coins to be counted.

The Measure of Everything

Measuring is essential as soon as a society starts to enclose and own land, or to trade in anything but the most basic objects, or to start building any but the simplest structures. Early civilizations needed to be able to measure distances, areas, volumes, and time. Some things that could be counted, such as grains of wheat, are more easily measured by volume, too. But units of measure didn't develop methodically. The jumbled mix of measures still in use in the UK and the United States is the legacy of earlier systems originating in Ancient Babylon, Egypt, and the Roman Empire with later Scandinavian, Celtic, Germanic, and Arab-influenced systems.

AT ARM'S LENGTH

Most of the earliest measurement systems, from China to pre-Columbian America, were based on dimensions of the human body or common objects, such as grains of wheat. Americans (and older British people) still measure short distances in feet, and a grain is still used as a unit of weight (it is the weight of a grain of barley, and has remained constant for over 1,000 years). The measure of gold and gemstones, the carat, has its origins in the carob seeds used originally by Arab jewelers to weigh

A tally man in an English hop field: pickers were paid by the bushel (a measurement of volume) and received half a tally stick by way of receipt.

precious metals and stones. The carob has seeds that are remarkably uniform in weight, making it ideal for measuring very valuable commodities.

The cubit, the unit of length familiar from the Old Testament in which Noah measured his ark, was an Egyptian measure equal to the distance from the elbow to the fingertips. It was subdivided into other units that also related to parts of the body:

AHEAD OF THEIR TIME

The inhabitants of the Indus Valley were among the first to produce a unified system of weights and measures and could measure distance, mass, and time with great precision. Their smallest measure, at 0.07 inches (1.7 mm), is the finest known from Bronze Age civilizations.

---

1 cubit = 28 digits
(a digit is a finger-width)

4 digits = 1 palm

5 digits = 1 hand

12 digits = 1 small span

14 digits = 1 large span

But the human body comes in all shapes and sizes, so one person's "hand" may be another's "palm." To overcome the obvious difficulties and potential for dispute, standard measures were needed. The cubit sticks used in Egypt

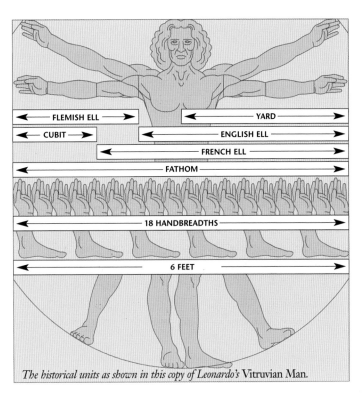

FLEMISH ELL

YARD

CUBIT

ENGLISH ELL

FRENCH ELL

FATHOM

18 HANDBREADTHS

6 FEET

The historical units as shown in this copy of Leonardo's Vitruvian Man.

were all copied from a royal standard made of black granite and measuring 524 mm (20.62 inches). The system successfully imposed uniformity. The Great Pyramid at Giza is built on a square base 440 by 440 cubits with variation of no more than 0.05 percent on any side—making it accurate to 4.5 inches (115 mm) in 756 ft (230.5 m).

ROMAN FEET

The foot divided into 12 inches originated with the Romans, though their foot was probably equivalent to 11.65 modern inches, or 296 mm. (There is some variation in the Roman foot, which appears to be deliberate but has never been fully explained.) They also had a palm, which was a quarter of a foot. Larger distances were

measured in furlongs (or *stade*), leagues, and miles. A furlong was an eighth of a mile, a mile was 5,000 feet, and a league was 7,500 feet. These measures, along with the Roman measures for weight based on pounds and ounces, spread through Europe and, hundreds of years later, were carried around the world.

BIG FEET AND SMALL FEET

During the centuries after the fall of Rome, measures developed and proliferated around Europe, but there was no uniformity. The length of a foot or the weight of a pound varied from place to place and sometimes according to what was being measured. So a gallon of wine contained 231 cubic inches, but a gallon of ale was

DEFINING THE PITCH OF A PITCHER

Chinese weights and measures developed independently of those in the West and the Middle East. The system was unique in incorporating acoustics in its standards. The standard vessel for measuring volumes of wine or grain was defined by the weight it could hold, its shape, and the pitch of the sound it made when struck. Two vessels of the same shape, material, and weight will only make the same sound if they hold equal volumes. The same word in Chinese is used for "wine bowl," "grain measure," and "bell."

China is going metric as trade increases with the rest of the world, but the old weights and measures system persists in much of the country.

282 cubic inches. (The first, known as the Queen Anne gallon, is still the standard gallon in the United States, though in the UK the gallon was redefined in 1824.)

Standardization came slowly, progressed by separate legislative acts in different countries. In the United States, the older English units survived after the UK had redefined them, resulting in the discrepancy between US customary and UK imperial units today.

WEIGHTS AND MONEY

It's no coincidence that the pound was the unit of both weight and currency in the UK for many centuries. When coins were made

"Uniformity of weights and measures, permanent, universal uniformity, adapted to the nature of things, to the physical organization and to the moral improvement of man, would be a blessing of such transcendent magnitude, that, if there existed upon earth a combination of power and will, adequate to accomplish the result by the energy of a single act, the being who should exercise it would be among the greatest of benefactors of the human race."

John Quincy Adams, American Secretary of State, 1821

The Romans introduced the pound, which has since been decimalized and may some day give way to the Euro.

from precious metals, weight was important, since weight and value were equivalent.

The Hebrew shekel was perhaps the earliest measure used for money and weight, and the Romans introduced the pound, which was then used in Europe for the next 2,000 years. In 1266, Henry III fixed the weight of a penny at 32 grains of wheat, with 20 pennies making an ounce and 12 (old) ounces to the pound. Eight pounds was the weight of a gallon of wine. Although the 12 and 20 switched places in the monetary system, with 12 pence to the shilling and 20 shillings to the pound, the equivalent of 240 pence to the pound—both sterling and avoirdupois—was established. The shilling has gone and the currency has been revised, but the legacy of the Roman pound and penny still survives in the British monetary system (though the Euro will probably replace them soon).

"FOR ALL PEOPLE, FOR ALL TIME"

The scientific community worldwide now uses SI units (*Système International d'Unités*) the seven standard metric units (gram, meter, Kelvin, ampere, candela, mole,

A NEAR MISS

An identical metric system to that eventually introduced in France was proposed in 1668 by Bishop John Wilkins, a founder of the Royal Society in England. In a long book on the possibility of an international language, he proposed an integrated system of measurement based on a decimal system and almost identical to the modern metric system. His unit of measurement was 997 millimeters—almost exactly a meter. The unit of volume was the equivalent of the liter. Wilkins' proposed system was never promoted and went largely ignored until rediscovered by Australian researcher Pat Naughtin in 2007.

STAR QUALITY OR "MAD AS A BADGER"?
Numbered scales can be used for qualitative comparisons. The star rating commonly used to grade hotels is a familiar and universal example of a qualitative scale. The badger rating for degrees of eccentricity used in parts of southern England is a localized system. Many websites invite users to rate preferences and experiences numerically. There is a whole science of evaluating the effectiveness of such rating systems.

and second). The metric system was first developed in 18th-century France. The need for a simpler, unified, and standard system of measures was pointed out by the vicar and mathematician Gabriel Mouton in

1670, but it took another 120 years before anything was done to provide it.

In 1790, Charles-Maurice de Tallyrand set the ball rolling again and the French Academy of Sciences recommended that a team determine the distance from the North Pole to the Equator, going through Paris. The first stage was to measure the distance on the meridian from Dunkirk in northern France to Barcelona in Spain. The day after King Louis XVI gave his approval, he was imprisoned by the French Revolutionary rulers. As a result, it was another year before the expedition started

Louis XVI gave his permission for the expedition—he was guillotined five months later.

A TIMELINE OF WEIGHTS AND MEASURES

ca.3000 BCE
Egyptians develop a royal standard for their basic measure of length, the cubit

ca.800
Holy Roman Emperor Charlemagne (r.768–814) tries to regulate weights and measures

1215
An English national standard for weights and measures is agreed and enshrined in the Magna Carta, the charter granted by King John (1199–1216)

1352
Edward III of England establishes that one stone equals 14 pounds, a value that remains to this day

1588
New standards issued by Elizabeth I in England (r.1558–1603)

ca.220 BCE
The first Chinese Emperor, Shi Huang Di (r.221–209/210 BCE) standardizes all weights and measures, even specifying the precise axle length to be used on carts

960
The first king of all England, King Edgar (r.957–975), decrees that weights and measures must accord with a standard kept in London

1266
Henry III fixes the relationship between money and weight in English currency, making one penny the weight of 32 grains of wheat and 240 pence to the pound

1496
New standards for weights and measures issued in England

A map of the heavens from the Harmonia Macrocosmica Atlas *by Cellarius (1660).*

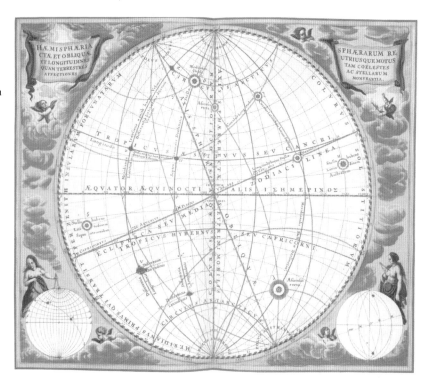

1670
Gabriel Mouton proposes a metric system of weights and measures in France

1790
George Washington's first message to Congress states the need for "uniformity in currency, weights, and measures"; Congress retains the English weights and measures system

1824
Redefinition of weights and measures in the UK, taking into account for the first time the conditions in which quantities are to be weighed and measured in establishing the standards

1878
The yard is redefined in the UK

1668
John Wilkins proposes a universal metric system of weights and measures in England

1707
A gallon of wine is fixed at 231 cubic inches. The measure had been used since the time of Edward I (r.1272–1307), but the act of 1707 fixed the size

1799
Standards of the metric system are defined in Paris, France

1866
The Metric Act allows the use of the metric system in the United States

1960
New *Système International d'Unités* (International System of Units or SI) formulated in Paris, France at the 11th General Conference on Weights and Measures

and even then it was beset with difficulties. War in France and Spain so hindered the project that it took six years to complete the journey. But in 1799 the metric system was formalized with two new units of measure, intended to be universal and enduring.

The meter was defined as "one ten-millionth part of a meridional quadrant of the Earth"; the gram was the mass of a cubic centimeter of pure water at 4°C (the temperature at which its density is greatest). A platinum cylinder, the Kilogram of the Archives, became the standard for the kilogram (1,000 grams).

The kilogram standard is now made of platinum-iridium alloy kept in Sèvres, near Paris; the kilogram is the only base unit still defined by a physical object. Attempts to find a better way of defining the kilogram are ongoing.

SILLY NUMBERS?

Calculating with the 1,760 yards in a mile, the 16 ounces in a pound, or the 160 square rods that make an acre has been the bane of many a schoolchild's life. The metric system looks simpler, based on multiples of ten and with clear relationships between measures of different quantities. But the SI units have some even more bizarre defining numbers. A meter is now the distance traveled by light in a vacuum in 1/299,792,458 of a second. And a second is the duration of 9,192,631,770 cycles of the radiation associated with a specific change in energy level of an atom of the isotope caesium-133.

In 1793 a meter was defined as 1/10,000,000 of the distance from the Pole to the Equator; now it's defined by the speed of light.

STONEHENGE

Stonehenge is a vast arrangement of concentric circles of stone and holes that were perhaps intended to hold posts or other stones near Salisbury in Wiltshire, England. The remains of the monument, built in three phases over a period of around 1,000 years between 3000 and 2000 BCE, consist of huge standing stones, some surmounted by stone lintels. The arrangement shows an ability to work with circles in space, and the curved lintel stones demonstrate an understanding of arcs of a circle—when all were in place, the lintel stones would have formed a true circle, not a series of straight stones. The only tools available to the builders were picks made of deer antlers and stone hammers, yet they were able to calculate and measure portions of a circle and distances. The northeast axis aligns with the position of the rising sun at the summer solstice, suggesting that some form of calendar had been developed.

Stonehenge is nearly as old as the pyramids if on a smaller scale. It, too, was built with a sophisticated understanding of geometry and the movement of the sun.

Early Geometry

Geometry deals with distances and angles, with lines, areas, and volumes. In its simplest and earliest forms, it works with lines and linear shapes in a flat plane. But from this it has been extended to dealing with curved lines in three-dimensional space and even to curved spaces in more dimensions that help us to explain the very fabric of the universe. On the way, it has given us architecture, astronomy, optics, perspective, cartography, ballistics, and much more.

PATTERN AND SYMMETRY: THE ESSENCE OF GEOMETRY

The earliest engagement with geometry pre-dates written number systems. Many early peoples have left evidence of their interest in repeated patterns, symmetries, and shape in the form of geometric patterns decorating their objects, structures, and dwellings.

Some of the earliest decorated objects have symmetrical patterns.

Some of these date from 25,000 BCE. Early structures built or aligned with considerable precision are further testimony to our ancestors' grasp of some simple form of geometry.

The annual flooding of the Nile, which erased boundary marks, was one of the prompts to the development of mathematics in Ancient Egypt.

PROBLEMS WITH LAND

Practical problems of geometry must have been tackled in building projects long before they were recorded in written form. The Sumerians, the Babylonians, and the Egyptians became quite adept at working with the geometry

Herodotus (ca.484–425 BCE) has been described as "the Father of History."

of two-dimensional shapes and three-dimensional objects, calculating distances, areas, and volumes. Documents from around 3100 BCE reveal that the Egyptians and Babylonians already had some mathematical rules for measuring storage containers, surveying land, and planning buildings. The Great Pyramid at Giza was constructed around 2650 BCE, demonstrating that the Egyptians already had a good grasp of geometry.

According to the Greek historian Herodotus, the Egyptians needed to be able to calculate areas because the seasonal flooding of the Nile swept away property boundaries. They needed benchmarks and surveying techniques to restore them properly. Egyptian geometers were sometimes referred to as "rope-stretchers" after their way of measuring and marking out distances and shapes using ropes. The

STRANGE GEOMETRIES

Enormous geometric patterns drawn in the Nazca Desert, Peru resolve into glyphs when seen from the air. They were created by the Nazca culture between 200 BCE and 600 CE. There are 70 individual figures, ranging in complexity from simple lines and geometric shapes to stylized animals, plants, and trees. Their significance is unknown, but their construction is evidence of some considerable skill with geometry among a people about whom we know little.

The very dry climate of the Nazca Desert has helped to preserve the giant geometric patterns drawn on the land around 2,000 years ago.

same technique no doubt served just as well to mark ground plans for building projects as to reclaim land that had been flooded.

WRITING IT DOWN

The earliest known mathematical document is the Ahmes papyrus (sometimes called the Rhind papyrus) from Egypt. It was written by the scribe Ahmes around 1650 BCE, copying from an older text written about 200 years previously, which itself may have contained much older material. It is a papyrus scroll 1 foot tall by 18 feet long (33 cm by over 5 m). It presents 84 mathematical problems, covering topics in arithmetic, algebra, and geometry, as well as weights and measures. Some of the problems are given a strictly practical presentation; for example, one asks, "A round field has diameter 9 khet. What is its area?" The so-called Moscow papyrus from around the same date includes instructions for working out the volume of part of a pyramid.

Because the Egyptians wrote on papyrus, which is fragile, very little of their mathematical writing survives. The people of Mesopotamia,

This Babylonian clay tablet featuring a problem in geometry is around 4,000 years old.

the fertile basin drained by the Tigris and Euphrates rivers, wrote instead on clay tablets, which they baked. These are much more enduring and over 100,000 survive. A Babylonian clay tablet dating from 1800–1650 BCE has been interpreted as presenting calculations for working out the hypotenuse of a right-angled triangle. The tablets also include methods for working with the areas of rectangles, triangles, and circles. One problem, for instance, asks about the distance the foot of a ladder moves if the top, leaning against a wall, slips down. They include an approximation for the square root of 2 which is accurate to five decimal places. The positional number system used by the Babylonians was better suited to all kinds of calculation than the Egyptian system—though it is hard to say whether an interest in numbers prompted the development of the better system or resulted from it.

Unlike the Egyptians, the Babylonians seem to have had a concept of general principles—that some mathematical statements will always be true in any situation of a given type. For example, one clay tablet shows the ratio of the sides to the diagonal of a square. Babylonian mathematicians had derived the ratio $1:\sqrt{2}$, the implication being that it is possible to find the diagonal of *any* square by multiplying its side by $\sqrt{2}$.

However, both the Egyptians and the Babylonians showed a cavalier disregard for accuracy. In some cases, their calculations give precise answers. In other cases, they use quite approximate methods for finding areas, but never concede that these areas are not accurate.

The area of this shape, for example,

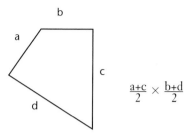

$$\frac{a+c}{2} \times \frac{b+d}{2}$$

they would calculate as a formula that would give only an approximate result.

THE BIRTH OF MATHEMATICS

While Egyptian and Babylonian mathematics often relate to particular, practical situations, a later civilization, the Ancient Greeks, took more of an interest in purely abstract problems. The ancestors of the Ancient Greeks began to enter the Greek peninsula from the north around 2000 BCE and were a force to be reckoned with by around 800 BCE. They ventured into Egypt and Mesopotamia, trading with and learning from their hosts.

There is no mention of Greek

MATHEMATICS IN VERSE

The earliest Indian texts to present mathematical problems are the Sulba sutras, Sanskrit texts that present problems and solutions relating to the construction and positioning of sacrificial altars. There were sutras—collections of aphorisms—on many different topics. The aphorisms are written in verse with prose commentaries and expositions. The sutras were originally passed on orally, the verse serving as an aid to memory. The Sulba sutras are among the oldest Hindu texts, the earliest dating from perhaps 800 BCE.

mathematics before the sixth century BCE, when the figures of first Thales and then suddenly Pythagoras appear. Thales of Miletus apparently brought Babylonian mathematics to Greece around 575 BCE. He has been called the "first mathematician" on account of having evolved a theorem and then demonstrated it, though whether he actually did this is impossible to say. What we know of Thales comes from later summaries of his reputation and we can't now tell how much of his mythic stature is deserved. The theorem that bears his name (the Theorem of Thales) states that any angle inscribed in a semicircle is a right angle (90°). This was known to the Babylonians around 1,000 years earlier, and Thales could have learned it in Mesopotamia. His demonstration of the theorem, if it existed, has not survived.

Writing around 900 years after Thales' death, Proclus (*ca.*410–485 CE) credits him with several fundamental geometric theorems:

• a circle is bisected by any diameter
• the base angles of an isosceles triangle are equal
• the opposite angles formed by two intersecting lines are equal
• two triangles are congruent (of equal

THALES OF MILETUS (*ca.*624–546 BCE)

Thales was one of the Seven Sages of Ancient Greece. He may have studied in Egypt as a young man and was almost certainly exposed to Egyptian mathematics and astronomy. If he wrote any works, they have not survived.

One story reported by Aristotle tells how he was able to predict a good harvest from observations of weather patterns and bought up all the olive presses in Miletus to prove how mathematics could make him rich. Diogenes Laertius reported that Thales was able to calculate the height of the pyramids by measuring their shadows, and he is said to have used his knowledge of geometry to determine the distance of a ship from the shore. He put his mathematical ability to military use, too. He is said to have predicted an eclipse which then led to a peaceful settlement in a war, and later to have helped King Croesus to get his army across a river by telling him to dig a diversion upstream to reduce the flow of the river until it was possible to ford it.

Thales is credited also with a cosmological model of the Earth as a vast disk floating in water. Ironically, Thales reportedly died of dehydration while watching a gymnastics contest.

shape and size) if two angles and a side are equal.

While Thales may be called the first mathematician, the title "father of mathematics" is often given to Pythagoras, who lived 50 years later. He is perhaps the best known of Greek mathematicians. No one can have come through school mathematics without learning Pythagoras' famous theorem: that in a right-angled triangle, the square on the hypotenuse is equal to the sum of the squares on the other two sides.

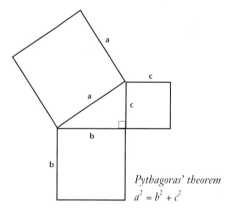

Pythagoras' theorem
$$a^2 = b^2 + c^2$$

> ### THE TETRACTYS
>
> For the Greeks, the number 10 was the most perfect number. They called it *tetractys* and revered it for being a triangular number, the sum of the digits 1 to 4, having as many primes as nonprimes before it, and being, in the words of Philolaus (died *ca.*390 BCE), "great, all-powerful, and all-producing, the beginning and the guide of the divine as of the terrestrial life."

It is likely, though, that the theorem was actually developed later by members of the Pythagorean school rather than by Pythagoras himself. As with Thales, no writings by him survive and we are dependent on later reports of his work and his reputation. (It is also possible the theorem was based on the breakthroughs of earlier mathematicians.)

The Pythagoreans were a secret brotherhood and held knowledge in common, so that individual attribution of work is now impossible. They took delight in the patterns and properties of numbers and sequences and believed that numbers were at the heart of all things. The group continued for many years after the death of Pythagoras.

THE NINE CHAPTERS

The earliest Chinese mathematical text, *The Nine Chapters on the Mathematical Art,* was first produced in the first century BCE. Many commentaries were written over the ensuing centuries, the best of which was by Liu Hui in 263 CE. The text demonstrates Pythagoras' theorem (derived independently) and shows how to calculate such distances as the height of a tower seen from a hill, the breadth of an estuary, the height of a pagoda, and the depth of a ravine. It also deals with finding areas and volumes of figures such as trapezoids, circles, segments of circles, cylinders, pyramids, and spheres.

PYTHAGORAS (*ca.*580–500 BCE)

Pythagoras was an Ionian (Greek) mathematician and philosopher. After traveling in the Middle East he moved to southern Italy around 532 BCE to escape the tyrannical ruler of his homeland, Samos. He is best known for the theorem that bears his name.

A contemporary of Buddha, Lao Tze, and Confucius, he established the Pythagorean Brotherhood at his academy in Croton. This was a religious and philosophical movement that influenced Aristotle and Plato and made an important contribution to the development of western philosophy. Pythagoras and his followers believed that everything was related to mathematics and everything could be predicted and measured in rhythmic patterns or cycles. The Pythagoreans were vegetarians as they believed in the transmigration of souls, and so any animal could house a former human soul. They also, rather curiously, believed beans to be special and would not eat them. Pythagoras is said to have been slaughtered by an angry mob when he refused to run through a bean field to escape their pursuit.

PYTHAGORAS CLAR OLYMP. 64
Pythagoras samius laudasse silentia fertur
Pythagore vera est numquit imago. tacet

THE GOLDEN AGE OF CLASSICAL GREECE

Athens in the fifth century BCE, between the Persian and Peloponnesian wars, saw one of the greatest flowerings of intellectual life in the history of the world. Sadly, no mathematical texts survive from the period and we have only a few scrappy accounts of the problems addressed by the great mathematicians of the day. Even so, we can deduce enough to see that mathematics was pursued for its own sake, for a delight in knowledge, and

"All things that can be known have number; for it is not possible that without number anything can be either conceived or known."

Philolaus, fourth century BCE

The Ancient Greeks were aware that the Earth is a sphere moving in space and believed that mathematics is the key to understanding the universe.

mathematicians made a distinction between the practical arithmetic of everyday life (of which nothing of theirs is recorded) and the higher pursuit of mathematics and logic, which has come down to us through the writings of those who benefited from their legacy.

THREE PROBLEMS

Greek mathematicians defined three great classical problems in geometry: squaring the circle, trisecting the angle, and doubling the cube, all to be achieved only with compass and straight edge. These problems were to tax mathematicians for 2,200 years until they were all proved to be impossible.

The issue of squaring the circle first appears in relation to Anaxagoras, a natural philosopher who wrote the first scientific best seller, *On Nature* (a copy could be bought in Athens for one drachma). Anaxagoras was imprisoned for denying that the sun was a deity, saying rather that it was a huge red-hot stone, bigger than the whole of the Greek peninsula and islands, and that reflected light from the sun illuminated the

because the Greeks believed that the workings of nature could be understood through mathematics. To them we owe the concept of the universe as a harmonious cosmos governed by laws that can be discovered by reason (rather than governed by an unknowable deity), the idea that the earth is a sphere and moves in space, and the concept of mathematical proof. The Greek

Anaxagoras (ca.500–428 BCE) was put in jail for denying the sun was a god.

76

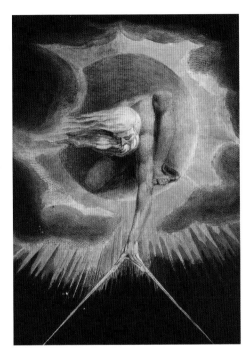

In The Ancient of Days, *poet and painter William Blake depicted God as architect of the universe, producing the Earth with geometric instruments.*

the population. The Greek craftsmen, who could not work out how to achieve what Apollo required, asked the philosopher Plato for advice. His answer was that the intention of the oracle was to shame the Greeks for their neglect of mathematics and geometry in particular. (There is another version of the origins of the problem in which Minos, king of Crete, commissions a tomb for his infant son, Glaucus, who died by falling into a vat of honey. Minos decides the tomb that is proposed is too small and demands that its size be doubled.) The Indian Vedic scriptures state that a second plea at an altar in the same place as a first plea demanded a cubic altar twice the volume of the first, and this may have suggested the problem to the Greeks.

The German mathematician Carl Friedrich Gauss stated that it was not possible to double the cube with only straight edge and compass, and this was proved by Pierre Wantzel in 1837.

moon. While in prison, Anaxagoras occupied himself by trying to discover a way of "squaring the circle"—given any circle, creating a square of exactly equal area using only straight edge and compass.

The problem of doubling the cube emerged at the time of the great plague in Athens (430 BCE). Eratosthenes reports that the people consulted the oracle at Delos, and the god Apollo demanded that to stop the plague they must double the size of his altar. They duly doubled the dimensions of the altar, but that of course increased its volume by a factor of eight (2^3), not two. Apollo was not satisfied, and the plague went on to kill around a quarter of

The plague of Athens is interpreted somewhat idealistically by Michael Sweerts (1618–64).

> *"In proceeding in [a mechanical] way, did not one lose irredeemably the best of geometry?"*
> Plato

Trisecting the angle is a less engaging problem and has no exciting mythical history attached to it; it is possible that it developed from the Egyptians' need to divide angles between stars in order to tell the time at night. The problem is simply to divide any angle into three equal parts using only straight edge and compass. It is possible to trisect some angles (a right-angle, for example) and there are mechanical methods for trisecting any angle which were known to the Greeks. However, the desire of the Greeks for pure or theoretical methods led them to continue the quest regardless of the lack of a practical need.

GEOMETRY RULES THE UNIVERSE

As we have seen, Greek mathematicians were reluctant to recognize irrational numbers (those that can't be expressed as a ratio of two whole numbers). Geometry cannot explain all things if it is limited to whole numbers and ratios of whole numbers—which becomes apparent as soon as we look at the diagonal of a square with a side of one unit. This alone would have been enough to bring the Pythagorean edifice crashing down. A second problem, expressed in the paradoxes of Zeno the Eleatic (*ca.*450 BCE), made matters worse.

ACHILLES AND THE TORTOISE

To demonstrate the absurdity of dealing in whole units, however small, Zeno proposed a race between Achilles and a tortoise. The tortoise has a head start, but even though Achilles can run very quickly he can never overtake the tortoise. In the time it takes Achilles to cover half the distance from the starting block to the tortoise, the tortoise has moved on. When Achilles covers half the remaining distance, the tortoise has moved on farther, though by a smaller amount. This goes on *ad infinitum,* so that Achilles is never able to come level with the tortoise.

Zeno's paradoxes show that, however much a unit of measurement is subdivided, it never expresses the continuum that we see in real life—even a sequence of infinitesimal steps is still artificial.

Confronting these two difficulties—the existence of irrational numbers and division into infinitesimal portions—forced a paradigm shift in Greek mathematics. At the time of Pythagoras, numbers were thought of as points, often represented concretely by pebbles (called *calculi*, giving us the word "calculation"). But by the time of Euclid, 200 years later, magnitude was represented by line segments—atomism had given way to continuity and the model of the basis of the universe had shifted from the discrete numbers of mathematics to the measurements of geometry. While $\sqrt{2}$ can't be represented as a number (in the Greeks' terms) it is very easy to draw it as a line segment.

DEMOCRITUS AND THE INFINITESIMAL

The chemist and philosopher Democritus (*ca.*460–370 BCE) proposed that everything is made up of infinitely small and varied particles moving around in empty space. The creation of our own and other worlds came about, he claimed, because the particles coagulated in different configurations, giving materials with certain similarities and differences. (The idea had already been suggested by Leucippus.) Extending this to geometric figures, a square pyramid, for example, can be seen as a stack of infinitely thin squares ranging from the largest at the base to the infinitely small at the apex. Because the layers are infinitely thin, each square is effectively the same size as its neighbors—but of course it can't be, for then the pyramid would be a cube.

Breaking down an area or volume into infinitesimally thin slices is the underlying principle of integral calculus, but Democritus could not progress toward this as a method because he could not get past his logical objections to the slices being different sizes. The method was used successfully by Antiphon, Eudoxus, and later by Archimedes, who derived the "method of exhaustion" to find the area of a shape (see page 146).

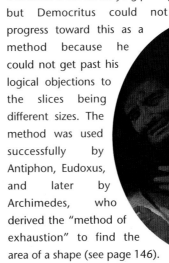

The odd couple: Democritus, the laughing philospher, pictured alongside Heraclitus, the crying philosopher.

BRINGING IT ALL TOGETHER

Practically no Greek mathematical texts are extant from before the fourth century, but we have not lost the work of this period. Perhaps the most famous mathematician of all time, Euclid of Alexandria, gathered together and recorded the inheritance of ancient geometry, codifying and extending it in his *Elements* around 300 BCE. By this time, the Greeks had discovered many of the standard curves (ellipse, parabola, hyperbola, and so on), a forerunner of integral calculus in the method of exhaustion, and methods for determining the volume of a cone and a sphere. Though Plato was not a mathematician himself, his academy in Athens was the center of the mathematical world and had helped to crystallize the distinction between pure mathematics and the practical application of numbers.

Euclid's *Elements* not only demonstrates the mathematics of the Ancient Greeks but also their development of logical method. Euclid presented five axioms and five

"common notions" and deduced from these several hundred theorems or proofs, exemplifying the principle of logical deduction which endured for centuries.

Although Euclid's text is famous for its treatment of planar geometry (the geometry of flat, two-dimensional shapes), it also deals with number theory, algebra, and solid geometry. It was intended as a textbook of elementary mathematics and does not deal with either simple arithmetic (which would have been beneath its intended readers) or the more complex geometry of curvilinear shapes and conics investigated later by Apollonius (which would go beyond what was required).

The five basic axioms from which Euclid develops everything else are:
1. Any two points can be joined by a single straight line.
2. Any finite straight line can be extended as a straight line.
3. A circle can be drawn through any center and with any radius.

The Oxyrhynchus papyruses—finds from an ancient town dump—contained the oldest and most complete diagrams from Euclid's Elements.

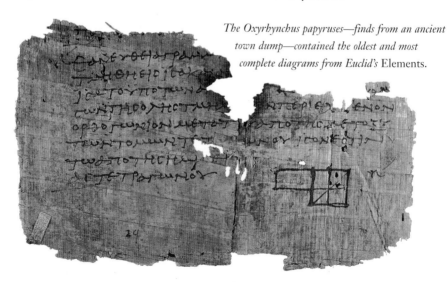

4. All right angles are equal to each other.

5. If two straight lines in a plane are crossed by another straight line (called the transversal), and the interior angles between the two lines and the transversal lying on one side of the transversal add up to less than two right angles, then the two straight lines can be extended until they eventually intersect on that side of the transversal.

> ### EUCLID OF ALEXANDRIA (*ca.*325–265 BCE)
>
> Euclid was a Greek mathematician who lived in Alexandria, Egypt, almost certainly during the reign of Ptolemy I (323–283 BCE). He is often considered to be the "father of geometry." His most popular work, *Elements*, is the most successful textbook in the history of mathematics and was used for over two thousand years. Euclid also wrote works on perspective, conic sections, and spherical geometry.
>
> Euclid would have written on papyrus scrolls and, as these decay readily, his work has come down to us only in copies. The oldest surviving version of Euclid's *Elements* is in a Byzantine manuscript written in 888 CE—making it closer in time to us than to Euclid! We can't be sure that we have Euclid's own text, rather than something improved upon or altered by later scholars.

The last of these is also called the "parallel postulate." It is not as self-evident and self-sufficient as the first four. Plato demanded that axioms should be simple, self-evident, and so clearly true that they need not be proven. While the first four meet his conditions, the fifth does not. This was probably evident even in Euclid's lifetime. However, it took until the 19th century to show that the final axiom could not be deduced from the others.

In addition, Euclid stated five "common notions" which are less strictly related to geometry:

1. Things that are equal to the same thing are also equal to one another.
2. If equals are added to equals, the wholes are equal.
3. If equals are subtracted from equals, the remainders are equal.
4. Things that coincide with one another are equal to one another.
5. The whole is greater than the part.

Euclid was writing just after the end of the Hellenic period, when both Alexander the Great and Aristotle had died. The empire of Alexander was broken up, and Athens lost its supremacy as an intellectual center, the intelligentsia convening instead in Alexandria in Egypt (Euclid included). Alexandria was the capital of Egypt and then fell under Roman rule when Cleopatra's army lost the battle of Actium in 31 BCE. The first to benefit from Euclid's work were the Romans, but mathematics was not highly regarded by Roman scholars and was taught for its practical usefulness rather than anything else. So an architect would need to understand geometry, calculation, and load-bearing and a merchant would need to understand arithmetic, but no one was particularly concerned to extend the boundaries of knowledge in mathematics for its own sake.

Built 70–80 CE, the Colosseum in Rome is an elliptical amphitheater and a masterpiece of Roman architecture and engineering.

The end of the Roman Empire in the West, when Germanic tribes under the leadership of Odoacer overran much of modern Italy, saw the end of mathematical activity in Europe for a long time. Instead, we must look to India and then the Middle East for developments in geometry as in other areas of mathematical endeavor.

HYPATIA OF ALEXANDRIA (CA.370–415)

Hypatia was the daughter of Theon of Alexandria, a notable mathematician and philosopher (it is from his version of Euclid's *Elements* that all surviving texts are derived). Hypatia was a Neoplatonist, and lectured on Neoplatonism and mathematics. She is the earliest known significant female mathematician. Sadly, none of her mathematical works survives, though her commentaries on the work of other mathematicians may be preserved in some of the annotations that have come down to us. She was murdered by a Christian mob in 415, incited by the patriarch of Alexandria to wipe out pagan scholarship. The Library at Alexandria was destroyed at the same time.

Trigonometry

Trigonometry is the branch of mathematics concerned with calculating angles, particularly in right-angled triangles. Until the 16th century, it was really a part of geometry, but since then it has come to be considered an independent area of mathematics.

As any polygon can be reduced to a number of triangles, trigonometry enables mathematicians to work with all areas or surfaces that are bounded by straight lines. Plane trigonometry deals with areas, angles, and distances in one plane. Spherical trigonometry deals with angles and distances in 3D space.

TRIANGLES INTO PYRAMIDS

The Egyptians had some knowledge of trigonometry as their building of the pyramids demonstrates. The Ahmes papyrus includes a problem that finds the *seked*, or slope, of a pyramid from the height and the base. It was expressed as the opposite ratio to our measure of gradient.

BELOW: Any polygon can be divided into triangles, which makes the calculation of area easy if you use trigonometry.

The Egyptians were not rigorous in their study of triangles, though. As in other areas of mathematics, they were interested in practical applications rather than pure trigonometry. Early Indian mathematicians, too, knew something of trigonometry. The Sulba sutras, in the context of describing altars, contain a calculation of the sine of $\pi/4$ (45°) as $\frac{1}{\sqrt{2}}$. However, it was left to the Greeks to develop trigonometry properly.

THE 360° TURN

The Greeks took the straight line and the circle as the basis of their geometry and from this developed trigonometry. The

ABOVE: We still show gradient as a ratio of the vertical rise and horizontal distance, though we've reversed the order since the Ancient Egyptians.

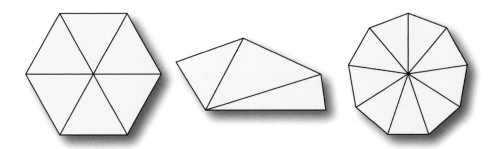

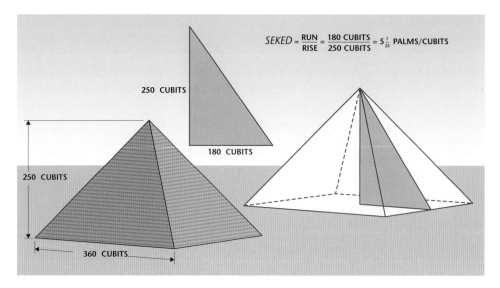

$$SEKED = \frac{RUN}{RISE} = \frac{180\ CUBITS}{250\ CUBITS} = 5\tfrac{1}{25}\ PALMS/CUBITS$$

250 CUBITS

180 CUBITS

250 CUBITS

360 CUBITS

convention of 360° in a circle and 60' in a degree originates with Hellenic math—it seems to have been in use by the time of Hipparchus of Bithynia (*ca*.190–120 BCE). It probably derives from the Babylonian astronomical division of the zodiac into 12 signs or 36 decans, and the annual seasonal cycle of approximately 360 days. The superior system used by the Babylonians for representing fractions

The Egyptians calculated the seked *or slope of a pyramid by imagining a right-angled triangle inside the structure.*

made it more useful than either the Egyptian or Greek systems, and Ptolemy (*ca*.90–168 CE) followed their base-60 system in dividing degrees into 60 minutes (*partes minutae primae*) and each minute into 60 seconds (*partes minutae secundae*).

SPHERICAL AND PLANAR TRIGONOMETRY

While a planar triangle is on a flat plane, a spherical triangle is one inscribed on the surface of a sphere. It is made up of arcs of three intersecting circles drawn around the sphere, or planes cutting through the sphere.

The first definition of a spherical triangle is found in a work by the

Zodiac clock: it was the Babylonians who divided the zodiac into 12 signs or 36 decans—reflecting their seasonal cycle of approximately 360 days.

The first mention of a spherical triangle comes in a work by the Egyptian Menelaus of Alexandria.

Egyptian Menelaus of Alexandria (*ca*.100 CE).

He developed the equivalents of Euclid's principles of planar trigonometry but applied to spherical triangles. Spherical triangles are clearly essential in astronomy and mapping.

While the angles of a planar triangle always add up to 180°, those of a spherical triangle always add up to more than 180°. There are other fundamental differences, too. Until around 1250, and the work of Nasir al-Din al-Tusi (1201–74), spherical trigonometry was always integrated with astronomy. Al-Tusi was the first to list six distinct types of right-angled triangles on a

spherical surface and the first to treat trigonometry as a discrete discipline. He developed spherical trigonometry to its current form.

THE RISE OF THE TRIANGLE

Hipparchus was the first to compile tables of trigonometric functions. His interest was in imaginary triangles "drawn" on the imagined sphere of the night sky that related heavenly bodies to one another so that he could calculate and predict the positions of planets. Hipparchus considered each triangle to be inscribed within a circle and developed a system of calculating angles from chords. He drew up tables of the chords produced by drawing angles of different sizes which relate to the modern concept of sines and cosines.

TRIGONOMETRIC FUNCTIONS

There are six trigonometric functions that enable us to calculate the size of an angle given two sides of a right-angled triangle, or the length of a side given one side and an angle.

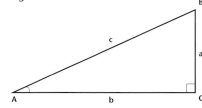

$$\sin A = \frac{a}{c} = \frac{\text{side opposite}}{\text{hypotenuse}}$$

$$\cos A = \frac{b}{c} = \frac{\text{side adjacent}}{\text{hypotenuse}}$$

$$\tan A = \frac{a}{b} = \frac{\text{side opposite}}{\text{side adjacent}}$$

$$\text{CSC (cosec) } A = \frac{c}{a} = \frac{\text{hypotenuse}}{\text{side opposite}}$$

$$\sec A = \frac{c}{b} = \frac{\text{hypotenuse}}{\text{side adjacent}}$$

$$\cot A = \frac{b}{a} = \frac{\text{side adjacent}}{\text{side opposite}}$$

In his astronomical text the *Almagest*, the Greek astronomer Claudius Ptolemy (100–170 CE) extended Hipparchus' work, deriving better trigonometric tables and loosely defining the inverse trigonometric functions arcsine and arccosine.

He used a nominal radius of 60 as the basis of his table of chords and gave values in steps of $1/2°$ from 0° to 180° accurate to 1/3600 of a unit. This is equivalent to a table of sines for every $1/4°$ from 0° to 90°. Ptolemy worked with Euclid's axioms and concentrated on planar triangles in order to develop his model of the heavenly bodies revolving around the Earth.

Ptolemy lived and worked in Alexandria. Details of his life have not been preserved—it's even possible that he may have been Greek in origin. His was the earliest work on trigonometry to be circulated in Europe in the Middle Ages and was used for many centuries. His model of the heavens survived intact until the work of the Polish

Hipparchus in his observatory at Alexandria, looking at the stars. He has been credited with the invention of the astrolabe as well as the armillary sphere.

astronomer Mikolaj Kopernik (Copernicus, 1473–1543), which put the sun at the center of the solar system.

SINE OF THE TIMES

After the Greeks, Indian and Arab mathematicians worked on trigonometry. Arab scholars translated and mastered the work of their Greek predecessors and soon went beyond it. The Indian mathematicians were largely working in their own tradition, which had drawn independently on the Egyptian and Babylonian heritage.

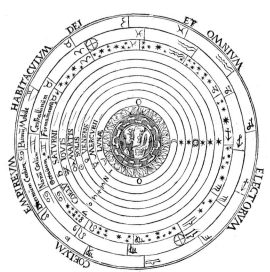

The geocentric (Earth-centered) universe (1539) shows Aristotle's four elements surrounded by the planets and the abode of God inter alia.

TRIANGLES AND WATER

One practical application of trigonometry was to calculate the gradient of water flow. The Sinhalese inhabitants of the city of Anuradhapura, Sri Lanka, used trigonometry for this purpose. Theirs was one of the greatest Asian civilizations of the ancient world. To farm the dry land around and supply water to the huge city, the Sinhalese built a highly sophisticated irrigation system, which consisted of overground and underground channels, reservoirs, and ponds.

In modern Anuradhapura, handpumps have replaced the complex irrigation system started 1900 years ago.

The Hindu mathematicians were the first to work with sines as we now define them. Early in the fourth century, or perhaps late in the fifth, the unknown author of the Indian astronomical treatise *Surya Siddhant*, calculated the sine function for intervals of 3.75° from 3.75° to 90°. The date of the text is not known—the surviving version may date from around 400 CE—but it claims to have been passed down directly by the sun god in 2,163,101 BCE! The *Aryabhatiya* by Aryabhata I (*ca*.475–550), which summarizes Hindu mathematics as they stood in the first half of the sixth century, includes a table of sines. Brahmagupta also published a table of sines for any angle in 628.

The first table of tangents and cotangents was constructed around 860 by the Persian astronomer Ahmad ibn 'Abdallah Habash al-Hasib al-Marwazi. The Syrian astronomer, Abu 'abd Allah Muhammad Ibn Jabir Ibn Sinan al-Battani al-Harrani as-Sabi' (*ca*.858–929), gave a rule for finding the elevation of the sun above the horizon

MEASURING THE EARTH

Eratosthenes (276–194 BCE) noticed that, while the sun is overhead at noon at the summer solstice in Syene (now Aswan), in Alexandria, 500 miles (800 km) northwest, it is at an angle of 7° at the same date and time. He assumed the sun's rays to be nearly parallel when they hit the Earth, since the sun is so far away. Working with trigonometry and the known distance between the two cities, he calculated the circumference of the Earth. The accuracy of his calculation can't be assessed exactly because the length of his unit of measure, the *stadia*, is not certain.

CALCULATING SINES

The sine function is the ratio of the side opposite an angle in a right-angled triangle to the hypotenuse.

To calculate the sine function, draw a circle of radius 1 and draw the required triangle within it, like this.

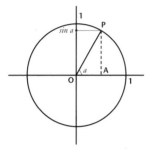

The distance OP, the hypotenuse, is the radius of the circle, 1. The y coordinate of point P gives the sine of angle a (= AP/1). The circle, called the unit circle (because it has radius 1) is, by convention, used to derive and relate all the trigonometric functions.

The Persian mathematician and astronomer Abu al-Wafa al-Buzjani (940–997/8) worked principally on trigonometry, but most of his work has been lost. He introduced the tangent function and improved methods of calculating trigonometry tables. He discovered the sine formula for spherical geometry:

$$\frac{\sin (A)}{\sin (a)} = \frac{\sin (B)}{\sin (b)} = \frac{\sin (C)}{\sin (c)}$$

A crater on the moon is named after him in honor of his extensive studies of the motions of the moon.

Arab mathematicians continued to refine tables and trigonometric exclusively in the service of astronomy until al-Tusi

by measuring a shadow (the principle on which sundials work). His "table of shadows" is effectively a table of cotangents for angles from 1° to 90°, at intervals of 1°. He also calculated the tilt of the Earth's axis, 23° 35'. It was through al-Battani's work that sines came to Europe and he may have discovered them independently of the work of Aryabhata.

Al-Battani calculated the tilt of the Earth's axis at 23° 35'. He also estimated the length of a solar year to be 365 days, 5 hours, 46 minutes, and 24 seconds.

Al-Tusi's pupil Qutb al-Din al-Shirazi was the first person to come up with a scientific explanation of the rainbow.

established trigonometry as a separate discipline in his observatory in Maragheh in the 13th century. One early development was the mathematical explanation of the rainbow by al-Tusi's pupil Qutb al-Din al-Shirazi (1236–1311). Ulugh Beg, the grandson of the great Mongol conqueror Timur (Tamberlaine the Great), established an observatory at Samarkand in the early 15th century and created tables of sines and tangents for every minute of arc, accurate to five sexagesimal places. It was one of the greatest achievements in mathematics up to this time.

Finding Directions

One spur to Arab advances in geometry and surveying was the need to determine the direction of Mecca (*qibla*) from any place, so that the devout Muslim could face the holy city for prayer as demanded by the Qu'ran. With this need in mind, Arab geometers adopted the stereoscopic projection, which produces a planar image of a spherical surface, with circles mapped either to circles or straight lines. This had first been used by Apollonius and Ptolemy.

From the ninth century, the Arabs perfected the astrolabe, an astronomical instrument originally designed in Ancient Greece. It consists of a series of concentric metal rings etched with the positions of the sun, moon, stars, and planets. Simply moving the rings replaced reams of tedious calculation. The astrolabe could be used for astronomy, timekeeping, surveying, navigation, and triangulation.

The combined Greek and Arab knowledge of triangles came to Europe with

The requirement for Muslims to pray to Mecca several times a day was a spur to improvements in finding directions.

the translation of many Arab texts into Latin from the 11th century. The Europeans took to the astrolabe enthusiastically and it remained the principal navigation instrument until the development of the sextant in the 18th century.

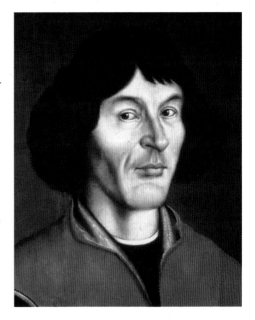

The sextant revolutionized navigation. It allowed sailors to plot their position by tracing the sun's course against the horizon.

INTO THE MODERN WORLD

Although medieval European scholars translated Arab and Greek works on trigonometry and other geometry, they added nothing new of their own. It was not until the explosion in scientific and mathematical knowledge in Europe from the Renaissance that trigonometry progressed again. Johannes Müller von Königsberg (1436–76), also known as Regiomontanus, was the author of the first book entirely devoted to trigonometry, *On Triangles of Every Kind*, printed in 1533. It brought together all the formulae required to work with planar and spherical trigonometry and was greatly admired and influential. His work was used and adapted by the great Polish astronomer Kopernik (Copernicus) in his new model of a solar system centered on the sun. Kopernik worked with the help of Prussian mathematician Georg Rheticus (1514–76). In his own work, Rheticus went further than Regiomontanus, finally making trigonometry about triangles. He discarded the old tradition of considering trigonometric functions with respect to the arc of a circle, freeing the triangle to stand alone. He calculated detailed tables for all

six trigonometric functions, and embarked on a set calculated to an even greater degree of accuracy but died before completing them. (they were finished by one of his pupils).

These developments came just before trigonometry, and geometry as a whole, took a new direction, becoming involved

In the 16th century, Mikolaj Kopernik demonstrated that the Earth circles the sun along with the other planets. This turned previous beliefs on their head.

DEADLY TRIANGLES

Galileo Galilei (1564–1642) discovered that the movement of a projectile is parabolic and could be separated into vertical and horizontal movement. This led to the formula for calculating the range of a cannon or other artillery weapon, disregarding air resistance:

$$\frac{V_0^2 \sin 2A}{g}$$

where g is the acceleration due to gravity (about 9.81 meters/second^2)

V_0 is the muzzle velocity (velocity at which the cannonball leaves the cannon, or bullet leaves the gun)

A is the angle of elevation

The maximum range is achieved when $A = 45°$.

The Inquisition forced Galileo to recant his belief that the Earth moved around the sun.

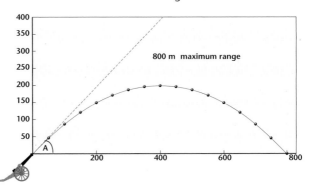

800 m maximum range

with algebra and the slow evolution of algebraic geometry (see Chapter 5). With this fundamental shift, trigonometry became more theoretical, separated from the real-world shapes with which it was originally concerned, and later even embroiled with imaginary and complex numbers.

At the same time, though, the practical applications of trigonometry were growing. The invention of accurate clocks, better navigation methods, and artillery, as well as new applications for optics and advances in astronomy, all demanded the application of trigonometry and aided its development in new directions.

MOVING ON

Triangles and circles are inextricably linked in the history of mathematics, and with Galileo's work on projectiles another curve, the parabola, becomes involved. Circles, curves, and the effects of revolving shapes in space to produce solids of revolution lift geometry away from straight lines on the flat page and move it into space. With the circles and curves, too, we begin to take steps toward contemplating infinity—that great bugbear of early mathematicians—which eventually would free geometry even from three dimensions to cavort through as many dimensions as we care to imagine.

CHAPTER 4

In the
ROUND

The world around us has provided the impetus for much of the development of mathematics. That the Earth itself is a sphere, and the sky looks like an inverted bowl above us, has put curves, circles, and spheres at the heart of geometry from early times. These features of the world have led to challenging problems with explaining, depicting, and modeling the universe as we experience it. How can we represent the three-dimensional environment we see in a flat drawing? How can we map the spherical Earth on a two-dimensional chart? Grappling with these issues led to a theoretical investigation that threw up further questions about dimension and geometry. Sometimes, the world does not seem to conform to the geometry laid out by Euclid, which was accepted for 2,000 years. New models for dealing with these situations have opened exciting and fruitful new avenues for mathematicians.

In the real world it seems as if parallel lines come together.

Curves, Circles, and Conics

The circle lies behind all trigonometry, as it defines a full revolution about a point. The triangle and the circle together formed the basis of astronomical geometry, with astronomical problems being explored by drawing imaginary triangles on the circular dome of the sky. Galileo's model of projectiles brings another curve to trigonometry and introduces a link between trigonometric functions and conics—curves that can be derived from slicing through a solid cone. In fact, triangles and curves have been inseparable since the earliest geometries. The trigonometric tables were all defined initially from triangles drawn within circles, using diameters and chords. Angles are measured with reference to the full revolution defined by a circle, which was divided at least by the time of Hipparchus into 360°.

THE MAGIC RATIO—π

From the earliest times, the circle has been endowed with religious and mystical significance. It is the perfect shape, having no sides (or infinitely many sides), the endless line, found everywhere in nature. People have known for thousands of years that the ratio between the diameter and circumference of a circle is always the same and have given this number special significance. We represent the ratio by the Greek letter π (pi), notation popularized by the Swiss mathematician Leonhard Euler (1707–83) in 1737, but first used by William Jones in 1706. Pi is an irrational number, and has an infinite number of decimal places (see table on page opposite).

Sir Isaac Newton, one of the most outstanding mathematicians of all time, calculated π to 16 places.

CALCULATING π

The Babylonians used an approximate value of the ratio we now call π, 3.125, which they obtained by calculating or measuring the perimeter of a hexagon drawn inside a circle.

The Ahmes papyrus shows a method from which a value of $^{256}/_{81}$, or about 3.16049, would be derived.

The Chinese text *The Nine Chapters* gives instructions for finding the area of a circle by squaring the diameter, dividing by 4, and multiplying by 3, so using a value of 3 for π.

Archimedes developed a more sophisticated method that involved drawing polygons both within and around a circle, giving upper and lower limits for a value of π. By adding more sides to the polygon, he could obtain increasingly precise limiting values. He settled on 96 sides, which gives a value for π between $^{223}/_{71}$ and $^{22}/_{7}$, or an average value of about 3.1418. It was Archimedes, too, who discovered that the same ratio can be used to calculate the area of a circle when multiplied by the square of the radius (πr^2).

Chinese, Indian, and Arab mathematicians calculated π to greater degrees of precision but had no better method than that of Archimedes. For example, in 263 CE Liu Hui used a polygon with 3,072 sides and obtained a value of 3.1416. At the end of the 17th century, better methods of calculation were developed. The English mathematician Sir Isaac Newton (1642–1727) used the binomial theorem to calculate π to 16 places.

Today, π has been calculated by computer to more than 10^{12} places and on personal computers there are plenty of programs for calculating π to a billion or more places. This degree of accuracy is completely unnecessary for most practical purposes. If the Earth's circumference is calculated from its radius using a value of π accurate to only ten decimal places, the result is accurate to around a fifth of a millimeter.

SQUARING THE CIRCLE

The problem of squaring the circle, although made famous by Anaxagoras, troubled earlier mathematicians, too. Ahmes gives a method for constructing a square of almost the same area as a circle, which consists of taking away one ninth of the diameter of the circle and using the remainder as the side of the square, though this is shown as a way of calculating the area of a circle, not solving the classical problem.

WHO	WHERE	WHEN	WHAT
Ahmes	Egypt	ca.1650 BCE	$^{256}/_{81}$ (3.16049)
Archimedes	Greece	ca.250 BCE	$^{223}/_{71} < \pi < ^{22}/_{7}$ (3.1418)
Chang Hong	China	130 CE	3.1622 ($\sqrt{10}$)
Ptolemy	Greece	ca.150	3.1416
Liu Hui	China	263	$^{3,927}/_{1,250}$ (3.1416)
Zu Chongzhi	China	480	$^{355}/_{113}$ (3.14159292)
Aryabhata	India	499	$^{62,832}/_{20,000}$ (3.1416)
al-Khwarizmi	Iran	ca.800	3.1416
Fibonacci	Italy	1220	3.141818
al-Kashi	Iran	ca.1430	3.14159265358979
François Viète	France	1593	3.1415926536
Adriaan van Roomen	Belgium	1593	3.141592653589793
Ludolph van Ceulen	Germany	1596	3.14159265358979323846264338327950290

(It is from this that we deduce the Egyptian value for π of 3.16049.)

We have already seen that the Greeks tried and failed to solve the problem geometrically. Later mathematicians also tried and all failed. Squaring the circle with compass and straight edge became such a preoccupation of both professional and amateur mathematicians in 18th-century Europe that in 1775 the Académie des Sciences in Paris passed a resolution saying that it would not look at any more proposed solutions. Soon afterward, the Royal Society in London did the same as they were inundated with faulty solutions. Some mathematicians even tried to fudge the issue by assigning a different value to π.

When Carl Louis Ferdinand von Lindemann (1852–1939) proved in 1880 that π is a transcendental number (i.e., not the root of any algebraic equation with rational coefficients), this demonstrated finally that squaring the circle is, in fact, impossible— it is quite impossible to work with a transcendental number using straight edge and compass.

CONIC SECTIONS

A circle is not the only curve. While the circle and circular arcs were the first curves to be studied and used, there are three other regular curves which came early to the attention of geometers. These are the parabola, hyperbola, and ellipse. Each can be formed by cutting through a cone. These are called conic sections.

The first influential work on conic sections was by Apollonius of Perga (*ca.*262–190 BCE), an Alexandrian-Greek geometer and astronomer known as "the Great Geometer." Although Apollonius wrote other works, only his treatise on conics has survived. The first sections draw on previous writings, but the later parts are

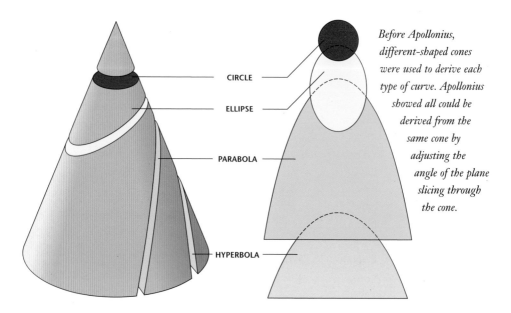

CIRCLE

ELLIPSE

PARABOLA

HYPERBOLA

Before Apollonius, different-shaped cones were used to derive each type of curve. Apollonius showed all could be derived from the same cone by adjusting the angle of the plane slicing through the cone.

BECOMING USEFUL

Apollonius was proud that his theoretical work was of value for its own sake—"They are worthy of acceptance for the sake of the demonstrations themselves"—and much of his work had little practical application in his day, but has since found uses in many areas of science. His work on the hyperbola produces a result equivalent to Boyle's law that defines the action of gases and his study of the tangents of an ellipse (though he did not know the term) is fundamental to understanding the movement of the planets and stars as well as in planning space travel.

More than 2,000 years after Apollonius lived, space travel has become a practical application of his work on curves.

completely original. Apollonius' work entirely replaced all work on conics that had come before as surely as Euclid's work had replaced all previous Greek geometries. Apollonius describes the derivation and definition of the curves he names and considers the shortest and longest straight lines that can be drawn from a given point or points on the curve. In this, he lays all the groundwork for the definition of curves by quadratic equations in the Cartesian coordinate system. Indeed, 1,800 years later, René Descartes tested his analytic geometry against a generalization of

Apollonius' theorem relating to a moving point and its relation to fixed lines.

Both the Arab and Renaissance mathematicians were heavily indebted to Apollonius. Though several Arab mathematicians studied conics, finding ways to calculate the areas and volumes of figures derived from conic sections, it was left for Omar Khayyam to take their study in a new direction. In using conics in his general solution of cubic equations, he anticipated Descartes to some degree, bringing geometry and algebra together (though he expressed a hope that his successors would be able to find algebraic solutions to finding roots). The rediscovery of Apollonius' work in the European Renaissance provided the groundwork for many of the advances in optics, astronomy, cartography, and other practical sciences.

The fabulous interior of the Hagia Sophia in Istanbul, formerly Constantinople: the altar is lit by sunlight at all hours of the day.

BEGINNING WITH OPTICS

In one of Apollonius' lost works he apparently discussed parabolic mirrors and demonstrated that light reflected off the inside of a sphere is not reflected to the center of the sphere. Optics was to become a major area for the application and development of work on curves. It could have most startling practical applications, too. In *ca*.200 BCE, Diocles demonstrated geometrically that rays of light that are parallel to the axis of a paraboloid of revolution (a solid produced by rotating a parabola) meet at the focus of the paraboloid. Archimedes is reported to have used this to set light to enemy ships from the shore. The focal properties of the ellipse were used by the architects of the Hagia Sophia Cathedral in Constantinople (537) to make sure that the altar was illuminated by sunlight at any hour of the day. Several Arab scientists investigated the properties of mirrors made from conic sections. Ibn al-Haytham found the point on a convex spherical mirror at which an observer would be able to see an object at a given position and showed how to design the curves needed for sundials.

The same properties can be applied to sound—the galleries in both the US Capitol and in St Paul's Cathedral in London are constructed so that a whisper uttered on one side of the gallery can be heard at the opposite point, but nowhere else. Even more recently, satellite dishes and solar collection dishes have used the reflective properties of a parabolic surface to focus the rays that strike them on to a central receiver or collector. In surgery the same geometry is exploited to focus ultrasound waves on organs or stones within the body.

Galileo's work on projectiles and Kepler's on planetary motion were among the earliest applications of conics to subjects other than optics. Kepler found that the Earth moves around the sun in an elliptical orbit, with the sun at one focus of the ellipse.

Later work on conics and curves used infinitesimal analysis to try to determine the area under curves or their length, but it was the invention of analytic geometry by Descartes and Fermat in the 17th century that paved the way for the modern definition of conics. Instead of deriving conic sections by cutting through a cone, the mathematicians of the 17th century and later defined them with algebraic equations, as the path traced out on a plane by points moving according to a second-degree equation in two variables. At this point in our story, conics disappear from geometry and re-emerge in algebra.

THE PERFECT PENDULUM

The Dutch scientist and mathematician Christiaan Huygens (1629–95) developed the pendulum clock after discovering a new curve, the cycloid. He discovered that a pendulum released from any height within a cycloid bowl will reach the bottom in precisely the same time—it does not matter how far it has to travel. Huygens went on to demonstrate properties of other curves. He used methods from analytic geometry and infinitesimal analysis to discover the lengths of curved lines and to become the first person to discover the surface area of part of a solid of revolution called a paraboloid, formed by rotating a parabola.

Solid Geometry

Solid geometry—the geometry of solid, three-dimensional objects—was needed as soon as humankind began building anything more complex than a simple hut (when trial and error rather than mathematics may have sufficed). One of the three problems facing classical mathematicians, the doubling of the cube, is a problem of solid geometry.

Problems in solid geometry relate to measuring the dimensions or volume of a three-dimensional shape. The volume measured need not be of a solid; it is likely that early uses for solid geometry related to measuring capacities as well as calculating the dimensions for buildings. Some of the problems in the Babylonian and Egyptian texts concerned calculating the volume of cellars and pyramids.

BASIC SHAPES

Plato identified five polyhedral solids with all faces the same. He associated these with the basic elements which he believed made up the physical world. These Platonic solids are the triangular pyramid (tetrahedron), cube (hexahedron), octahedron, dodecahedron, and icosahedron. Plato claimed that earth was made of cubic particles, fire of pyramids, air of octahedrons, and water of icosahedrons. He claimed, "… the god used [the dodecahedron] for arranging the constellations on the whole heaven." In his *Elements*, Euclid gives a thorough account of the Platonic solids and repeats Plato's assertion that there are only five regular solids.

Before starting work on a pyramid, Egyptian builders had to calculate its volume in order to acquire the right amount of stone.

| TETRAHEDRON | HEXAHEDRON OR CUBE | OCTAHEDRON | DODECAHEDRON | ICOSAHEDRON |

The German astronomer Johannes Kepler (1571–1630) tried to associate the Platonic solids with the known planets and formed a model of the solar system in which the solids were nested within one another. Although he had to give up the model, he did, in the process of working on it, discover two regular stellated polyhedra in 1619. These are formed by extending the edges, or faces, of polyhedra until they meet, forming new shapes. Louis Poinsot discovered a further two in 1809. In 1812, Augustin Cauchy proved that there were no more regular star polyhedra.

Platonic solids: these were the basic elements that Plato believed made up the physical world. He further believed earth was made of cubic particles, fire of pyramids, air of octahedrons, and water of icosahedrons.

Although Plato is credited with first describing the Platonic solids, they are all represented on carved stone balls 4,000 years old found in Scotland. At least one of Kepler's polyhedra was known before he wrote about it, too. A stellated polygon is depicted on the marble floor of the Basilica of San Marco in Venice, Italy, which dates from the 15th century.

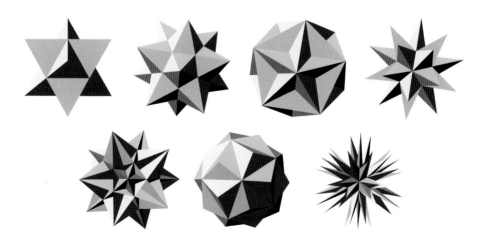

Stellated regular and irregular polyhedra are created by extending the faces of a polyhedron until they intersect. Some polyhedra produce many stellations, others very few.

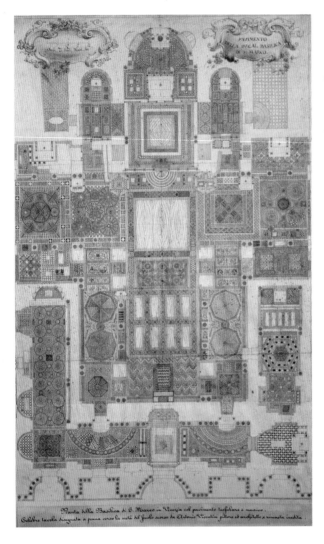

The groundplan for the Basilica of San Marco, Venice: a stellated polygon is depicted on the marble floor dating from the 15th century.

MEASURING VOLUME

Just as a two-dimensional polygon can be reduced to a series of triangles, so a three-dimensional polyhedron can often be reduced to regular solids for the purposes of calculating volume. Methods for calculating the volume of a cube, square, or triangular pyramid, cylinder, and cone were known to the Ancient Egyptians. But the volume of shapes that cannot be reduced to any of these is harder to calculate. Archimedes is credited with realizing that the volume of an irregular shape can be found by measuring the volume of water it displaces, a discovery that reportedly led him to leap naked from his bath and run down the street shouting, "Eureka!"

THE GOLDEN CROWN

The Roman writer Vitruvius (died *ca.*25 BCE) told a story in which King Hieron commissioned a solid gold crown and asked Archimedes to determine whether the crown the jeweler made was really solid gold. Clearly, Archimedes could not damage the crown to test it. He realized while in the bath that he could measure the water displaced by immersing the crown, then weigh the crown and calculate its density. By comparing this with the known density of gold he could work out whether or not it had been adulterated with a cheaper, less dense metal.

While most famous for his "Eureka!" moment in the bath, Archimedes also explained the principles of the lever, the device upon which mechanics is based.

A sphere is not usually considered a regular solid as it has no angles, edges, or faces. Archimedes proved that the volume and surface area of a sphere are two-thirds that of a cylinder the same height and diameter. The earliest demonstration of the volume of a sphere, $\frac{4}{3}\pi r^3$, was by the Chinese mathematician Zu Chongzhi (429–500).

VOLUMES AND CALCULUS

After establishing mathematical methods for discovering the volumes of regular polyhedra, or solids that can be broken into regular polyhedra, and a practical method for measuring the volume of irregular solids (by immersion), there was little left to calculate but the volume of irregular solids and solid conics. These problems were not solved until the invention of calculus at the end of the 17th century.

Zu Chongzhi was the first to measure the volume of a sphere, but also created a new calendar system, as commemorated in this statue in Shanghai, China.

Seeing the World

For all the theoretical purity of the Greek mathematicians, math comes from and impacts on our relations with the real world. The developments that had begun with an interest in rarefied logic, distanced from real-world applications, led in Renaissance Europe to a rich cross-fertilization between the arts, sciences, and mathematics that resulted in new ways of seeing the world. These, in turn, led back into new mathematical ideas.

The way we see the world—indeed, the universe—around us interested and inspired geometers for centuries. Not only the mechanics of how we see and how light behaves, but the difficulties of representing and modeling what we see have both benefited from and prompted developments in geometry. Perspective geometry is the study of the relationship between figures and how they are mapped or represented, and began with the study of shadows cast by objects and the way items in the distance appear to the eye.

PUTTING THINGS IN PERSPECTIVE

The Arab scientist and mathematician Abu Ali al-Hasan ibn al-Haytham (*ca*.965–1040) worked with geometry to formulate his ideas on optics. He developed some of Euclid's work, redefining parallel lines, and used conic sections to help in his exploration of the reflection and refraction of light. He arrived at the accurate model of light rays emanating from an object rather than being sent out by the observer's eye (which was the model adopted by some scientists). He described a pyramid of rays coming from the object, some of which reach the eye of the observer. He went on to determine the point of reflection from a plane or curved surface, using conic sections. The work of al-Haytham came to the West through Latin translation and prompted one of the greatest revolutionary events in the history of art—the discovery of linear perspective in the Italian Renaissance.

It was the Florentine architect and engineer Filippo Brunelleschi (1377–1446) who first rediscovered

The Dead Christ *by Andrea Mantegna (1431–1506) is a superb early example of the application of the principles of perspective in western art.*

the architectural principle of linear perspective that had been known to the Greeks and Romans. Brunelleschi demonstrated the principle of perspective in two illustrative panels that have been lost, but in 1435 his work was incorporated in *Della pittura* (On Painting) by Leon Battista Alberti (1404–72).

Alberti suggests that a painting is like a projection of an image on to a vertical plane cutting through the pyramid of rays of light at some point between the object (the apex of the pyramid) and the observer's eye. The painting includes a "point at infinity" (now called a vanishing point) at which parallel lines in the painting converge.

MAPPING THE WORLD

Recording larger-scale images of the world required a different type of application of geometry. Surveyors used trigonometry for the new method of triangulation that made accurate maps possible for the first time. Triangulation was first suggested in Europe by the Flemish mathematician Gemma Frisius (1508–55) in 1533, though crude versions of triangulation were used in

The dome of the Cathedral of Santa Maria del Fiore in Florence, Italy, (1420–36) is the pinnacle of Brunelleschi's achievement; it is held up by pressure from the weight of timbers.

Members of the 1874 geographical survey conduct triangulation work at the top of Sultan Mountain in San Jan County, Colorado, USA.

Ancient Egypt and Greece, and Heron of Alexandria described a primitive theodolite in the first century CE.

From each end of a base line, angles of sight to a distant object are measured using a theodolite. Trigonometric methods then give the distance to the object. By covering a geographic surface with measured and calculated triangles, the entire area can be mapped. The first large-scale mapping project was carried out by Willebrord van

A Renaissance world map based on the writings of Ptolemy's Geography. *The art of cartography developed dramatically during the age of discoveries.*

PTOLEMY AND THE AMERICAS

Though Ptolemy's most famous work was the *Almagest,* he also wrote a *Geography,* which remained influential for over a thousand years. He developed two projections and introduced lines of latitude and longitude, though the inaccuracy of measurements led to considerable errors in his longitudes. He also overestimated the extent of the Earth's surface covered by the Hellenic lands and consequently his calculated size of the Earth was smaller than the real thing.

The earliest surviving European maps from the Middle Ages are heavily reliant on Ptolemy's *Geography.* When explorers planned to sail to India by heading west they would have expected the journey to be much shorter than it actually was. Perhaps if Columbus had realized the true nature of the undertaking he would not have attempted the voyage that led him to the Americas.

A romanticized artist's impression of Columbus landing in America in 1492. He must have been relieved to strike land after a longer-than-expected journey.

Roijen Snell (1581–1626), who surveyed a stretch of 80 miles (130 km) in Holland with 33 triangles. The French government decided to survey the whole of France, which took more than a hundred years to complete. The British surveyed all of India between 1800 and 1912, discovering Mount Everest in the process.

From the mid-15th century onward, explorers were discovering and charting new lands, beginning with the Portuguese exploring the African coast. While surveying deals in straight lines, the cartographers who were aiming to record the large expanses of newly discovered lands needed a way of representing in two dimensions terrains that are actually draped over the surface of a sphere. The method Ptolemy had used in his *Geography* (rediscovered in Renaissance Europe) did not work for the enlarged world. Instead, cartographers adopted the stereographic projection that astronomers used to portray the sky. But, of course, that depicts the

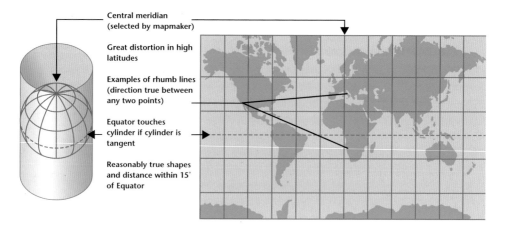

Central meridian
(selected by mapmaker)

Great distortion in high
latitudes

Examples of rhumb lines
(direction true between
any two points)

Equator touches
cylinder if cylinder is
tangent

Reasonably true shapes
and distance within 15°
of Equator

interior of a hemisphere and the cartographers needed to represent the exterior of a sphere. (A stereographic projection projects a sphere on to a flat plane from a projection point which is then not visible on the map. Areas near to the projection point are distorted.)

The most successful variant developed was the Mercator projection, made first by the Flemish map maker Gerardus Mercator (1512–94). He drew the Earth as though projected on to a cylinder tangential with the equator. Parallels and meridians are drawn as straight lines spaced to produce an accurate ratio of latitude to longitude at any point on the map. When the cylinder is unrolled, the flat map is revealed. Although

A map of the world using the Mercator projection, showing how it is derived from the projection of the globe on to a cylinder.

the projection was useful for navigation, it distorts areas particularly near the poles. A Mercator projection of the Earth shows Greenland as approximately the same size as Africa, for example, whereas in fact the area of Africa is around 14 times that of Greenland.

AND BACK TO MATHEMATICS...

The intense discussion of perspective and projections fed back into mathematics, stimulating discussion of the properties of perspective in general. The most significant

---

PROJECTIVE GEOMETRY

Projective geometry formalizes the principle central to linear perspective in art of showing parallel lines meeting at a point that represents infinity. It rejects Euclid's fifth axiom (the parallel postulate). Desargues extended the convenient trick of perspective drawing, taking it off the artist's page and formulating a space in which parallel lines actually do meet at infinity. He used this projection to study geometric figures, including conics.

outcome was in the work of Girard Desargues (1591–1661) which eventually led to the development of a more rigorous projective geometry in the 19th century.

Desargues was a French mathematician, architect, and artist, a friend of both René Descartes and Pierre de Fermat (1601–65), the leading mathematicians of his day. He developed a geometric method for constructing perspective images of objects and wrote a highly theoretical text explaining the geometry of constructing perspectives in 1636. The engraver Abraham Bosse restated Desargues' work in 1648 in a more accessible form, presenting what is now known as Desargues' theorem. This states that if two triangles are situated in three-dimensional space so that they can be seen in perspective from one point, then corresponding sides of the triangles can be extended so that they intersect, with the points of intersection all lying on a line. This works as long as no two corresponding sides are parallel. A modification of the theorem takes account of this. Desargues' work was popular for around 50 years, and was read by Pascal (see page 43) and Leibniz, but was then largely ignored until it was rediscovered and published again in 1864. Both Desargues and Pascal studied the properties of figures that were preserved and those that were distorted by different methods of projection. For example, a flat map of a spherical world cannot accurately represent both distance and shapes.

Gerardus Mercator holding his globe.

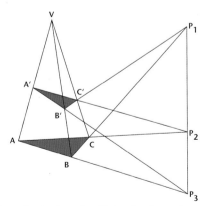

From the viewing point, V, the triangles are in perspective. If corresponding edges of the triangles are extended until they meet (BC and B'C' etc), the points of intersection, P1, P2, and P3, lie in a straight line.

In addition to his work on projective geometry, Poncelet is regarded as the most influential engineer in history, proving the work-kinetic energy theorem.

The principles of projective geometry were rediscovered by Jean-Victor Poncelet (1788–1867) in the early 19th century. Poncelet had been left for dead at Krasnoy, Russia, in 1812 after fighting in Napoleon's Russian campaign. He was then captured and imprisoned at Saratov, and worked on problems of perspective and conics while in prison. His solution to the need for a modification to Desargues' theorem in the case of parallel sides was to change the nature of Euclidean space. Poncelet postulated points at infinity, each line having a point at infinity and parallel lines having a point in common at infinity. This became the basis of the new projective geometry. Poncelet ignored geometric measurements of distances and angles in order to find other properties of figures which do not vary when they are projected. These involved collinear points—points that fall on a line in the original also fall on a line in the projection—and some special ratios between distances. Projective geometry could be used to further work on conics (since all conic sections can be seen as projections of a circle).

Other Worlds

Euclidean geometry provides us with the tools we need for working with the geometry of planes—but perfectly flat planes exist in only small or ideal environments. We live on a spherical Earth, in a universe with at least three physical dimensions. In representing the curved surface of the Earth, or of the sky as it appears to us, on a flat piece of paper we are necessarily distorting it. Some of these problems are addressed by projective geometry. However, as we move away from the perfect, regular curvature of a sphere, more problems of geometry relating to curved surfaces emerge. Though the ancients were aware of difficulties in marrying Euclid's geometry to curved surfaces, it was not until the 19th century that mathematicians developed new models to address them.

SPHERICAL GEOMETRY

The first non-Euclidean geometry to develop, spherical geometry, tackles measurements on the surface of a sphere. It is the geometry of the surface of a sphere.

> *"It has been demonstrated by mathematics that the surface of the land and water is in its entirety a sphere... and that any plane that passes through the center makes at its surface, that is, at the surface of the Earth and of the sky, great circles."* Ptolemy, *Geography*, ca.150 CE

Using spherical geometry, we are able to measure distances on planets and moons with some degree of accuracy.

One anomaly of spherical geometry is immediately apparent. A line in spherical geometry is the shortest path between two points, just as it is in the geometry of flat planes, but it looks very different. A line drawn across the surface of a sphere, if continued long enough, meets its own beginning, becoming a circle with its center at the center of the sphere. This is called a geodesic, or great circle—so a straight line becomes a circle! All other aspects of planar geometry are then adapted accordingly—so angles are defined between great circles, for instance.

A line on a spherical surface is defined not by its length but by the angle under which its endpoints appear when viewed from the center of the sphere. This angle is called the *arc angle*. It is usually measured in *radians*. The arc angle multiplied by the radius of the sphere gives the length of the line over the surface.

Some differences between planar and spherical geometry quickly become obvious. On the surface of a sphere, we can define a closed shape using only two lines (or great circles)—think of a segment of an orange. Clearly, we can't make a shape from only two straight lines in a plane. Spherical triangles have other special properties. The angles always add up to more than 180°; how much more than 180° is determined by the size of the triangle and is called the spherical excess (E). This can be used to calculate the area of the triangle:

$$\text{area} = E \times r^2$$

where r is the radius of the sphere and E is given in radians. This is called Girard's Theorem after the French mathematician Albert Girard (1595–1632).

RADIANS AND DEGREES

One radian = $\frac{180}{\pi}$ degrees; there are 2π radians in a circle, or π radians on a straight line. Radians were first used as a measure of angles by the English mathematician Roger Coates in 1713. He recognized that the radian is a more natural unit of measure than degrees, though he did not use the name. The term first appears in print in an exam paper set at the Queens College, Belfast, Northern Ireland in 1873.

The outer surface of an orange segment is a shape bounded by only two straight lines.

Early astronomers and surveyors were working with spheres as they looked at the sky and the Earth. They became aware early on of difficulties with Euclidean geometries when applied to spheres. However, it took many centuries for the possibility of alternative geometric rules to be accepted.

ELLIPTIC AND HYPERBOLIC GEOMETRIES

Curved surfaces give rise to two non-Euclidean geometries. Lines drawn on curved surfaces do not behave in the same way as those drawn on a plane, as we have seen with spherical geometry. Most importantly, Euclid's fifth postulate does not hold. In Euclidean planar geometry, two lines both drawn perpendicular to a given line, L, will be parallel. For a curved surface this is not true. On an elliptical surface, there are no such lines—two lines drawn perpendicular to a third line will eventually intersect. A perfectly elliptical surface is a sphere and spherical geometry is a special— the simplest—model of elliptical geometry.

On a hyperbolic surface, two lines drawn perpendicular to L will diverge. If the curvature is exactly right, the hyperbolic surface will be the inside of a sphere, but

Three simple diagrams representing the behavior of lines with a common perpendicular in each of the three types of geometry.

> "The hypothesis of the acute angle is absolutely false; because it is repugnant to the nature of straight lines."
>
> Saccheri

otherwise it will be some vast bowl. Clearly, the reverse of a hyperbolic surface is elliptical—the outside of a sphere is elliptical and the inside of the sphere is hyperbolic.

REJECTING ALTERNATIVE GEOMETRIES

That the behavior of lines on a curved surface is contrary to Euclid's rules of geometry disturbed mathematicians. For many centuries, they tried to deny all non-Euclidean geometries. The Italian mathematician Giovanni Girolamo Saccheri (1667–1733) tried to prove that they could not exist, but ended up doing the

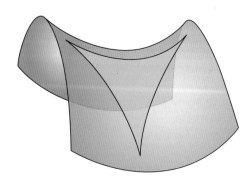

A triangle drawn on the hyperbolic surface of a saddle demonstrates that the angles inside a triangle in hyperbolic geometry can add up to less than 180°.

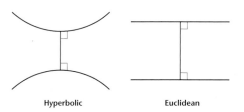

Hyperbolic   Euclidean

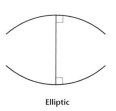

Elliptic

János Bolyai's compasses in the Bolyai Museum,
Marosvásárhely, Romania (where he died).

opposite, demonstrating the possibility of
alternative geometries and deriving some of
the principles of hyperbolic geometry. His
work apparently drew on writings of the
Iranian mathematician Omar Khayyam
(1048–1131), though he may have
developed his arguments independently.

Saccheri took as his starting point a
parallelogram proposed by Omar Khayyam.
The parallelogram is formed from a pair of
parallel lines, with sides drawn between
them, perpendicular to the two lines. (In
normal planar geometry this looks like a
regular rectangle.) He then considered three
possibilities: that the internal angles are 90°,
less than 90° (acute), or more than 90°
(obtuse). Although it looks pretty obvious
that they are 90°, his aim was to prove that
they could not be anything else, and so
support the fifth postulate. It turned out that
these alternate hypotheses were not as
absurd as Saccheri had hoped. His reasoning
for refusing the other two possibilities
was not sufficiently rigorous and,
though he rejected them, he did
not successfully disprove the case
for acute angles. It emerged later
that the case of the acute angle

gives a system equivalent to hyperbolic
geometry and that of the obtuse angle gives
elliptical geometry. Saccheri's work had little
impact in his lifetime, and its importance was
not recognized until Eugenio Beltrami
rediscovered it in the mid-19th century.

DAWNING ACCEPTANCE

Hyperbolic geometry re-emerged with the
independent work of the Hungarian János
Bolyai (1802–60) and the Russian Nikolai
Ivanovich Lobachevsky (1792–1856) around
1830. Bolyai published in German and
Lobachevsky in Russian; it was not until
Lobachevsky published in German, too, that
his work came to wider attention. The great
German mathematician Carl Friedrich
Gauss claimed to Bolyai that he had already
discovered most of what Bolyai revealed

A statue of János Bolyai and
his father, Farkas, also a well-
known mathematician, who
instructed his son from an
early age.

CARL FRIEDRICH GAUSS (1777–1855)

Carl Friedrich Gauss was a child prodigy, born to uneducated, impoverished parents in Germany. He had an amazing capacity for mental arithmetic and claimed to be able to calculate logarithms in his head more quickly than he could look them up in a table.

Gauss made great advances in mathematics and its applications in astronomy, statistics, earth sciences, and surveying. He produced several important theorems and proofs in many areas, and his work on curvature underpinned Einstein's theory of relativity. Working with his physics professor Wilhelm Weber, he studied the earth's magnetic field, developing methods that were still used until the second half of the 20th century. The pair also constructed the first electromagnetic telegraph in 1833. Gauss's claim to have worked out hyperbolic geometry before Bolyai placed a strain on relations between the two, since Bolyai considered that Gauss was trying to steal his ideas. In fact, Gauss's personal diaries suggest that on several occasions he came up with ideas many years and even decades before others published them, but did nothing with them himself.

The Gauss-Weber monument in Göttingen, created by Carl Ferdinand Hartzer in 1899.

when he published in 1832, but had not himself publicized it. This was possibly true, and both Lobachevsky and Bolyai had links with Gauss that could have given them insights into thoughts contained in his teaching and correspondence. In this case, Gauss would have been the first to develop a consistent, non-Euclidean geometry. The work of Lobachevsky and Bolyai had little impact until Gauss's ideas were published after his death in 1855.

Gauss suggested treating hyperbolic and elliptical

A portrait of Nikolai Ivanovich Lobachevsky. He spent most of his career as a professor at Kazan University.

surfaces as "spaces," since although they exist in three dimensions they actually have only two dimensions and only two variables are needed to specify a point on them. He showed that a surface could be described entirely with reference to distances and angles measured on it, and without giving information about its placement in three-dimensional space.

RIEMANN AND IRREGULAR CURVES

Although Bolyai and Lobachevsky had demonstrated that a set of alternative methods for working with hyperbolic surfaces was feasible, there was no model equivalent to Euclid's planes, lines, and points for dealing with the geometry of curved surfaces. Such a model was provided by the Italian Eugenio Beltrami (1835–1900) in 1868. Importantly, he demonstrated that hyperbolic geometry was consistent if Euclidean geometry is consistent. Beltrami developed spatial models which are now called the pseudosphere, Poincaré disk, Klein model, and the Poincaré half-plane.

On the Poincaré disk, distances at the edges are larger than distances near the center, though this is not apparent as the disk curves away from the viewer. In Escher's picture, "Circle Limit III," the figures are the same size all over the surface. The way that the Mercator map projection distorts the size of countries near the poles is similar—Greenland looks larger than it is, for example—but on a Poincaré disk the distortion is the other way, with distances seeming smaller than they are. The shortest distance between two points on the edge of a Poincaré disk is an arc of a circle drawn at right-angles from the boundary of the disk.

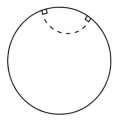

Similarly, the center of a hyperbolic circle is not in its middle:

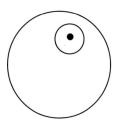

A computer graphic showing the curvature of space created by a black hole. The curvature of space-time was established by Einstein's theory of relativity.

115

German Bernhard Riemann (1826–66) extended hyperbolic geometry to work with surfaces that do not have uniform curvature. He developed a system for describing the curvature at any point on a surface in space using only ten numbers. Riemann geometry involves postulating higher dimensions (i.e., dimensions beyond the three familiar, physical dimensions). He began with a concept of *n*-dimensional space, and used calculus to provide geodesics for any curved surface. His work underpins much of modern physics, including Einstein's theory of relativity.

The attempts to prove non-Euclidean geometries led to greater rigor in scrutinizing the *Elements*. The German mathematician Moritz Pasch (1843–1930) saw the need for concepts, axioms, and logical deductions based on these axioms to underpin the new geometries as well as old mathematics. This contributed to the drive, led by David Hilbert at the start of the 20th century, to axiomatize all of mathematics and provide a firm foundation in proof for even the most seemingly obvious deductions (see page 199).

INSIDE OUT?

Curved surfaces form the basis of the branch of mathematics called topology. It became one of the most important areas of development in mathematics in the middle of the 20th century (1925–1975). Although, as Gauss and Riemann showed, they exist in *n*-dimensional space, they have only two dimensions of their own. Surfaces can even

> *"The views of space and time which I wish to lay before you have sprung from the soil of experimental physics, and therein lies their strength. They are radical. Henceforth space by itself, and time by itself, are doomed to fade away into mere shadows, and only a kind of union of the two will preserve an independent reality."*
>
> Hermann Minkowski, 1908

be twisted into shapes that appear to be three-dimensional objects, producing curious anomalies that have no distinct inside or outside.

The simplest example of this idea is the Möbius strip, easily made by taking a strip of paper, twisting it once, and gluing the ends together. The strip has only one side—you can run a finger over the whole surface, both sides of the original paper strip, in a continuous movement.

The Klein bottle is an extension of this principle, requiring a further dimension. Although the bottle is necessarily drawn intersecting its own surface, as it is modeled

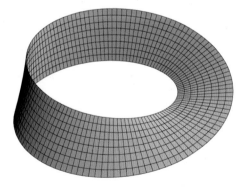

A Möbius strip. The visual deception created by such shapes—is it two- or three-dimensional?—forms the basis of the work of artist M.C. Escher (1898–1972).

EINSTEIN, RIEMANN, AND THE SPACE-TIME CONTINUUM

The general theory of relativity postulated by Albert Einstein (1879–1955) uses the concepts of Riemann geometry and an extra dimension, making a four-dimensional space called space-time. (Space-time was first suggested by Hermann Minkowski (1864–1909), following the publication of Einstein's special theory of relativity in 1905.) In general relativity, space-time is curved, with the degree of curvature increasing close to massive bodies. Curvature is the result of the interaction of mass-energy and momentum producing the phenomenon we know as gravity. Thus Einstein's theory replaces the "force" of gravity familiar from Newtonian mechanics with multidimensional, non-Euclidean geometry.

Following Minkowski's lead, Albert Einstein added a fourth dimension to the worlds of mathematics and geometry with his theory of space-time.

This curvature, and the principle of relativity, was proved in 1919 by observations of an eclipse. Einstein predicted that light rays would be distorted by the curvature of space produced by the gravity of a nearby star or planet.

At the moment of an eclipse, a star would appear to be in slightly the wrong place because of this distortion. Measurements made by Sir Arthur Eddington (1882–1944) in Principe Island, Gulf of Guinea, proved that this was indeed the case.

FLATLAND

The novella *Flatland: A Romance of Many Dimensions,* written and illustrated by Edwin Abbott Abbott in 1884, satirized the social hierarchy of Victorian Britain in a mathematical tale. The narrator, a square, occupies a two-dimensional world, Flatland. He dreams that he visits a one-dimensional world, Lineland, but cannot convince the ruler that life in two dimensions is possible. The square is visited by a sphere, but can't conceive of a three-dimensional world until he visits it. The square then tries to convince the sphere that more dimensions might exist, but he won't be persuaded. It becomes a criminal offence to suggest in Flatland that a three-dimensional world is possible.

In another dream, the square is introduced to Pointland and again has no success persuading the ruler there of the existence of alternative worlds.

Title-page illustration from Flatland, *showing all the places the square visits in his dream.*

mathematically it is curved through four dimensions and has no intersection. The inside becomes the outside seamlessly. A Klein bottle can be dissected to give two Möbius strips.

The Dutch artist M. C. Escher drew several scenes that played with the ideas of impossible surfaces and structures. The Penrose triangle, first drawn by the Swedish artist Oscar Reutersvärd in 1934, was popularized by the mathematician Roger Penrose in the 1950s. He called it "impossibility in its purest form."

TRIVIA

The name "Klein bottle" is a misinterpretation of the German *Kleinsche Fläche* ("Klein surface"), taken as *Kleinsche Flasche* ("Klein bottle"). The name has stuck (even in German) and several glass-blowers have made literal "Klein bottles," though necessarily with an intersection. There is a display of these in London's Science Museum.

MOVING ON

Impossible geometries, it turned out, are not so impossible after all, and the fact that we can't visualize something doesn't mean that it can't exist. As with mapping three-dimensional space to flat planes, all that is needed is a consistent and

The impossible triangle in East Perth, Western Australia. The structure is actually disjointed at the top, and has been photographed from one of the two spots from which it is designed to be seen.

A mathematician named Klein
Thought the Möbius band was divine.
Said he: "If you glue
The edges of two,
You'll get a weird bottle like mine."

Anonymous limerick

thorough method. The way these representations work, especially for spaces in more than three dimensions, is by way of a coordinate system. This can be explored and manipulated mathematically using algebra. Algebra and geometry developed in parallel, with considerable cross-fertilization, until the 17th century. Then the remarkable work of two Frenchmen in that century brought them together and provided the tools needed for Riemann and other non-Euclidean geometries to be formulated.

The Penrose triangle.

CHAPTER 5

The
MAGIC
FORMULA

Algebra is familiar to most people in the form of equations that must be solved, either equations set as exercises at school or equations formed to model problems in economics, science, or some other discipline.

The representation of unknown quantities by symbols, which is fundamental to algebra, evolved slowly. Although Ancient Egyptian and Sumerian mathematicians dealt with problems that involved unknown quantities, they did not express them in the form of equations as we do now. Indeed, not until the late 16th century did the familiar form of an equation evolve. We now have many ways of solving equations, including the use of graphs. This has been made possible by the crowning achievement of René Descartes, who brought together geometry and algebra in the system of Cartesian coordinates, which allows an equation to be plotted as a graph.

At this angle, the Tower of Babel defies the rules of God and geometry.

Algebra in the Ancient World

It is impossible to disentangle simple algebra from geometry, for it was in problems relating to two- and three-dimensional geometry that algebraic questions first surfaced. Early on, specific, practical problems in algebra were neither systematized nor represented in a way that we would now recognize as algebra—yet they provide the origins of algebra as it was later formulated.

FIELDS AND CELLARS

Babylonian clay tablets in the British Museum include a number of problems that would now be formulated as quadratic or cubic equations. These relate to building projects and involve working with areas and volumes.

Some problems related to dividing up an area in parts with different proportions. It is easy to see how a problem in area can lead to a quadratic equation.

The method for solving simultaneous equations named after Carl Friedrich Gauss had been used in the East 2,000 years earlier.

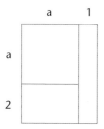

Here the area of the larger (enclosing) rectangle is

$$(a + 2)(a + 1) = a^2 + 3a + 2$$

Similarly, cubic equations can be derived from Babylonian problems relating to digging cellars. The earliest known attempt to write and tackle cubic equations is in the form of 36 problems about construction in a clay tablet nearly 4,000 years old. Such problems were expressed in words by both the Babylonians and Egyptians, and by mathematicians for many centuries afterward—for example "the length of a room is the same as its width plus 1 cubit; its height is the same as its length less 1 cubit."

The Babylonians did not adopt any form of notation to express or treat problems of these types, although they may have had general methods, or algorithms, that they used to solve comparable problems. The Ancient Egyptians, too, solved practical problems that would now be expressed as linear or quadratic equations, but again without recourse to any formal notation and without recognizing them as equations.

The Chinese text *The Nine Chapters* (second–first century BCE) includes a chapter on solving simultaneous linear equations for two to seven unknowns. They were solved using a counting board or surface and could include negative coefficients. The description of equations with negative coefficients is the earliest known use of negative numbers. The method used is now known in the West as Gaussian elimination after Carl Friedrich Gauss, who used it 2,000 years later.

FROM GEOMETRY TOWARD ALGEBRA

In the middle of the third century CE, the Hellenistic mathematician Diophantus of Alexandria developed new methods for solving problems that would now be shown as linear and quadratic equations. His work, *Arithmetica* (of which only part has survived), contains a number of algebraic equations and methods for solving them. Diophantus applied his methods to the problems in hand, but did not extend them to general solutions. Like the earlier Greeks, he dismissed any solutions that were less than zero, and when an equation

yielded more than one solution he stopped after arriving at the first—even if there were an infinite number of solutions (as for an equation of the type $x - y = 3$).

He developed a method for representing equations that was less cumbersome than writing them out in words, but was still not comparable with modern methods. As the Greeks used the letters of their alphabet for numbers, there were no recognizable symbols immediately available to represent variables. We can use x, y, a, b, m, n, and so on to stand for variables and constants because we have separate symbols for numbers, and so an expression such as 2x is unambiguous. Diophantus adopted some variants on Greek letters, and used symbols to indicate squaring and cubing. His system of abbreviations was an intermediary stage between the purely discursive explanation of problems and the purely symbolic in use now. It also gave him the opportunity, not seen or exploited before, of dealing in higher powers than cubes. Some of his problems include a notation that means "square-square" or "cube-cube," indicating powers of 4 and 6 respectively.

In addition, Diophantus had no concept of an equality—of two balanced expressions between which parts could be moved or on which identical operations could be carried out. Nor did Diophantus deal with more than one unknown at a time. He always sought a way to convert a second unknown into an expression built around the first. So, for example, in a problem that calls for two numbers whose

---

INDIAN QUADRATICS

An ancient Indian text, one of the Sulba sutras written by Baudhayana around the eighth century BCE, first cites and then solves quadratic equations of the form $ax^2 = c$ and $ax^2 + bx = c$. These occurred in the context of building altars, and so relate to a practical problem in three dimensions.

sum is 20 and the sum of whose squares is 208, Diophantus would not write, as we may, $x + y = 20$; $x^2 + y^2 = 208$, but might term them $(x + 10)$ and $(x - 10)$, the second equation then becoming $(x + 10)^2 + (x - 10)^2 = 208$.

DIOPHANTINE EQUATIONS

Diophantine equations are those in which all the numbers involved, including those in the solutions, are whole numbers (which can be positive or negative). They fall into three categories: those with no solution, those with a fixed number of solutions and those with infinitely many solutions.

For example, the equation

$$2x + 2y = 1$$

has no solutions, because there are no values for x and y that are whole numbers that can give the answer 1 (the sum of two even numbers is always even).

The equation $x - y = 7$ has infinitely many solutions as we can continue to pick larger and larger values of x and y.

The equation $4x = 8$ has only one solution: $x = 2$.

Diophantine equations are useful for dealing with quantities of objects that cannot be divided—such as numbers of people. So, for instance, if there is a choice of cars to take 24 people on a trip, some of which carry four and some of which carry six passengers, and all must be full, we could write a Diophantine equation, since the only useful solutions assign whole numbers of people to whole numbers of cars:

> **ORDERS OF EQUATION**
>
> **Polynomial equations** are those that contain a series of terms, each of which has a variable raised to any power, multiplied by a constant (ordinary number). For example in the following equation
>
> $$x^2 + 2x - 8 = 0$$
>
> the first term consists of $x^2 \times 1$, the second of $x^1 \times 2$ and the last the constant -8 (or $x^0 \times -8$). Mathematicians refer to polynomial equations as being of the first order, second order, and so on, depending on the highest power they contain.
>
> So a quadratic equation such as that above is called a second-order equation; an equation including a cubed term (x^3) is a third-order equation.

$$4x + 6y = 24$$

(This has the additional requirement that the values of x and y must both be positive.) Math problems of the following type use Diophantine equations: "A boy has spent 96 cents on sweets and bought 4 chocolate mice, 2 lollipops, and a chocolate bar. What is the cost of each item?"

Diophantine equations of the form

$$ax + by = c$$

are linear equations (a graph drawn of the equation would be a straight line). Another Diophantine equation,

$$x^2 + y^2 = z^2$$

relates to Pythagoras' Theorem and produces Pythagorean triplets (e.g., 3, 4, 5: 9 + 16 = 25).

Although Diophantine equations are named after Diophantus, he was not the first to work on them. The Indian Sulba sutras deal with several Diophantine equations. However, Diophantus differed markedly from earlier Indian and Babylonian mathematicians in that his problems were purely theoretical—he was not concerned with building altars, digging cellars, or taxing grain, and his numbers do not relate to quantities in the real world. He was also concerned only with precise answers using whole numbers. It is probably for this last reason that there are few cubic equations in Diophantus' *Arithmetica*. Although the questions that Diophantus deals with may not look unusually difficult, his approach was genuinely innovative and has had a lasting effect on later mathematicians. Indeed, it was while trying to generalize a problem raised by Diophantus, to divide a square into two squares, that Fermat arrived at his famous Last Theorem (see page 140).

GOING BEYOND THE CUBE

While Diophantus had a form of notation for powers greater than three, he did not make any great use of it. Another Alexandrine, Pappus of Alexandria, approached the issue, but again did not come to grips with it. He was the first to state clearly that linear, or first-order, algebraic problems relate to a single line or one dimension; second-order problems relate to two dimensions or areas, so are planar, and third-order problems relate to three dimensions or volumes, so are solid. Investigating the properties of curves defined by lines in planes and volumes, he came up against the possibility of equations of a higher order. However, he dismissed it since "there is not anything contained by more than three dimensions." Diophantus was too much wedded to algebra and Pappus to geometry for either of them to make the conceptual leap into algebraic geometry, though they both approached the jumping-off point. It was one of Pappus' geometric problems of lines and loci that eventually led Descartes to invent algebraic geometry in the 17th century.

OBSERVATIO DOMINI PETRI DE FERMAT

Cubum autem in duos cubos, aut quadratoquadratum in duos quadratoquadratos & generaliter nullam in infinitum ultra quadratum potestatem in duos eiusdem nominis fas est diuidere cuius rei demonstrationem mirabilem sane detexi. Hanc marginis exiguitas non caperet.

Translation of Fermat's Last Theorem: *It is impossible for a cube to be the sum of two cubes, a fourth power to be the sum of two fourth powers, or in general for any number that is a power greater than the second to be the sum of two like powers. I have discovered a truly marvelous demonstration of this proposition that this margin is too narrow to contain.*

AL-MAMUN'S DREAM

The caliph al-Mamun (786–833) is said to have had a dream in which Aristotle appeared to him. As a consequence, the caliph ordered translations to be made of all the Greek texts that could be found. The Arabs had an uneasy peace with the Byzantine empire and negotiated the acquisition of texts through a series of treaties. Under al-Mamun's caliphate and at his House of Wisdom, complete versions of Euclid's *Elements* and Ptolemy's *Almagest* were translated, among others.

A 15th-century painting of Aristotle. Al-Mamun's reign was noted for his huge efforts in the translation of Greek philosophy and science.

The Birth of Algebra

With the development of the Indo-Arabic number system and the adoption of zero, something approaching modern algebra became possible. The Arab mathematicians, in drawing together the best of Indian and Greek mathematics and extending it, laid the foundations of a proper algebraic system and even gave us the term "algebra." They found algebra more appealing than the Greeks had done and there were also spurs to its development within their own society. The incredibly complex laws of inheritance, for example, made the calculation of proportions and fractions a tedious necessity. On top of that, the constant need to find the direction of Mecca made algebra, like geometry, a tool worth developing.

AL JABR WA-L-MUQABALA

The word "algebra" is derived from the title of a treatise written by the Persian mathematician and member of the House of Wisdom, Muhammad ibn Musa al-Khwarizmi, called *Al-Kitab al-Jabr wa'l-Muqabala* (The Compendious Book on Calculation by Completion and Balancing). This presented systematic methods for solving linear and quadratic equations. The modern word "algorithm" comes from the name "al-Khwarizmi," too. In his book he gives methods for solving equations of the types $ax^2 = bx$, $ax^2 = c$, $bx = c$, $ax^2 + bx = c$, $ax^2 + c = bx$, and $bx + c = ax^2$ (in modern notation). Like Diophantus, he only considered whole numbers in equations and their solutions; he had the additional requirement that the numbers must also be

Omar Khayyam was also responsible for the reform of the Persian calendar. His Jalali calendar is the basis of that still in use today in Iran and Afghanistan.

positive, while Diophantus allowed negative numbers. Al-Khwarizmi wrote out all problems and solutions in words and had no symbolic notation. Ironically, since his work is credited with introducing Hindu-Arabic numerals to Europe, he even wrote the numbers out in full.

After showing how to tackle equations, al-Khwarizmi went on to use Euclid's work to provide demonstrations using geometry. Euclid's propositions were entirely geometric, and al-Khwarizmi was the first to apply them to quadratic equations. The method he developed, of systematizing the

GHIYAS AD-DIN ABU AL-FATH OMAR IBN IBRAHIM KHAYYAM NISHABURI (1048–1131)

Omar Khayyam was a mathematician, astronomer, and poet born in Iran, probably to a family of tent-makers. He lived most of his life on a modest pension provided by a friend who became grand vizier to the Seljukid empire. His *Treatise on Demonstration of Problems of Algebra* (1070) set out the basic principles of algebra and was responsible for the transmission of the Arab work on algebra to Europe. He worked on the triangular arrangement of numbers known as Pascal's triangle and is sometimes considered the originator of algebraic geometry, which uses geometry to find solutions to algebraic equations.

A 19th-century English translation of Omar Khayyam's collection of four-line poems, the Rubayat. *Many Persian scholars were also poets.*

$$(a+b)^n = a^n + na^{n-1}b + \frac{n(n-1)a^{n-2}b^2}{1 \times 2} + \frac{n(n-1)(n-2)a^{n-3}b^3}{1 \times 2 \times 3} + \frac{n(n-1)(n-2)(n-3)a^{n-4}b^4}{1 \times 2 \times 3 \times 4} + \ldots + \frac{nab^{n-1}}{1} + b^n$$

cases and then applying a geometrical solution, was adopted by later Arab mathematicians and perfected by Omar Khayyam (see below). Al-Khwarizmi's work stands for algebra as Euclid's *Elements* did for geometry, and remained the clearest and best elementary treatment until modern times.

Omar Khayyam followed a similar procedure to al-Khwarizmi, using Greek geometric work on conic sections to demonstrate his solutions to cubic (third-order) equations. Omar Khayyam produced general solutions for cubic equations where the Indian mathematicians had worked only with specific cases. In 13th-century China, Zhu Shijie developed solutions for cubic equations without reference to Omar Khayyam's work.

SHAPES, NUMBERS, AND EQUATIONS

In Pascal's triangle, each number is the sum of the two numbers above it. The pattern forms the binomial coefficient series. In Iran, it is called Khayyam's triangle and in China Yang Hui's triangle after the Chinese mathematician Yang Hui (1238–98) who also worked on it.

```
          1
        1   1
      1   2   1
    1   3   3   1
  1   4   6   4   1
```

Before Omar Khayyam wrote on Pascal's triangle, it

The equation shows how to find the coefficients and variables for any expanded binomial expression of the form $(a + b)^n$.

had been studied in India by Pingala (fifth–third century BCE), though only fragments of his work survive in a later commentary. Another Arab mathematician, Abu Bakr ibn Muhammad ibn al Husayn al-Karaji (*ca*.953–1029), had also worked on it and is credited with being the first to derive the binomial theorem (see above):

The Indian mathematician Bhattotpala (*ca*.1068) wrote out the triangle up to row 16.

The triangle provides a quick way of expanding expressions such as $(x + y)^3$, since all that is needed is to take the coefficients from (in this case) line 3 (since it is a third-order equation), giving the result:

$1x^3 + 3x^2y + 3xy^2 + 1y^3$.

MOVING AWAY FROM AREAS

Although geometry provided good methods of proving algebraic solutions, it was as algebra moved away from the restrictions of real-world geometry that the idea of an abstract equation, relating to numbers rather than measures or quantities, became

"Whoever thinks algebra is a trick in obtaining unknowns has thought it in vain. No attention should be paid to the fact that algebra and geometry are different in appearance. Algebras are geometric facts that are proved."

Omar Khayyam

128

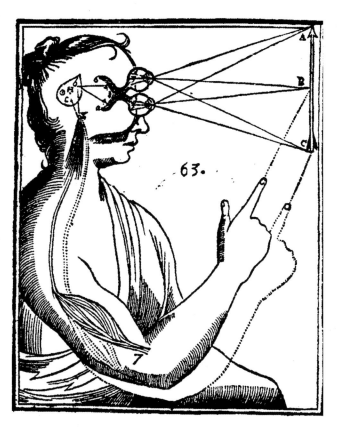

An illustration from Descartes'
The World, *in which he set down
his theories on light, the senses,
biology, and many other topics.*

possible. The Arab mathematicians were willing to treat commensurable and incommensurable numbers alongside one another, and to mix magnitudes in different dimensions, both of which the Greeks were unwilling to do.

Combined with the Hindu-Arabic number system and the acceptance of zero, this allowed algebra to move forward and away from its roots in practical geometry. When Omar Khayyam and al-Khwarizmi had recourse to geometry to demonstrate their algebraic results, they were not imagining their algebraic problems in terms of lengths, areas, and volumes but using geometry theoretically as a tool to represent algebraic problems.

This relationship between the two, developed over the next 500 years, resulted eventually in the analytic geometry of Descartes and Fermat.

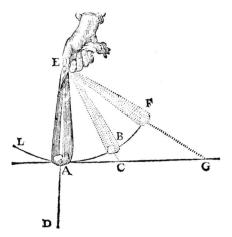

An illustration showing the movement of objects, from Descartes'
Principles of Philosophy.

Writing Equations

Omar Khayyam died in 1131 and already Arab mathematics was in decline. Scholars from the Arab world were to make few further contributions in the field. Luckily, at the same time that political and religious groups were fracturing the Arab cultural world, the intellectual spirit was reawakening in Europe. During the 12th century Gerard of Cremona translated 87 works of Greek and Arab scholarship into Latin, working at Toledo. These included Ptolemy's *Almagest*, Euclid's *Elements*, and al-Khwarizmi's *Algebra*. In England, Robert of Chester translated al-Khwarizmi in 1145 and Adelard of Bath translated Euclid's *Elements* in 1142. After centuries spent recovering and consolidating earlier learning, European mathematicians began to make their own contribution to the development of algebra. Germany was the focus of these new developments in the 16th century. Perhaps the most important of the new German works on algebra was *Arithmetica integra* by Michael Stifel (*ca*.1487–1567). He allowed the use of negative coefficients in quadratic equations and as a consequence reduced the various types of quadratic to a single form. He used negative powers to denote reciprocals, too, giving

$$2^{-1} = 1/2^1 = {}^1/_2, \ 2^{-2} = 1/2^2 = {}^1/_4$$

and so on. Even so, he did not allow negative roots in equations and referred to negative numbers as *numeri absurdi*. He was similarly distrustful of irrational numbers, which he said are "hidden under some sort of cloud of infinitude." He proposed using a single letter to denote an unknown quantity, repeating the letter for powers of the number—so if c is the unknown, cc is c^2 and ccc is c^3.

TOWARD A NOTATION FOR EQUATIONS

Algebra without the symbols we use now was cumbersome and long-winded. Yet the modern notation is a late arrival on the scene. In Italy, the symbols $\tilde{p}$ and $\tilde{m}$ came to be used for plus and minus as abbreviations for the words *più* (more) and *meno* (less). But Latin was full of abbreviations for words and groups of letters that are written repeatedly and this was not particularly original. The introduction of arithmetical operators—symbols showing the type of computation to carry out—did not begin until the late 15th century.

The first symbols to be used were + and –, though originally they were to show a surplus and a deficit in warehouse quantities. They soon took on their modern role as arithmetic operators. They were first printed in a book by Johannes Widmann (born *ca*.1460), one of several German mathematicians who published on algebra in the late 15th and early 16th centuries.

Even after the development of symbols, many mathematicians continued to follow the rhetorical model, writing out the problems they were posing and solving as discursive text with little or no recourse to symbolic abbreviation (syncopation). Although Western math did not have a thorough and consistent symbolic algebra until the 17th century, the western part of the Islamic world used symbolic notation in 14th-century commentaries intended for teaching.

ROBERT RECORDE (1510–58)

Robert Recorde was born in Wales and taught mathematics at the Universities of Oxford and Cambridge. He trained in medicine and was private physician to Edward VI and then Mary I. He was also Controller of the Royal Mint.

Recorde re-established mathematics in England, when the country had not seen a notable mathematician for 200 years. He explained everything in careful detail, in steps that were easy to follow and in English, as he wanted to make mathematics as accessible as possible. Most of his works were written in the form of dialogues between a master and a student. In 1551 he published an abridged version of Euclid's *Elements*, making the text available in English for the first time. He first used the equals sign, though using much longer lines than we do now. It took 100 years before the sign was universally accepted above alternative notations.

In 1558 Recorde was imprisoned for failing to pay £1,000 libel charges. He died in prison the same year.

IN MEMORY OF
ROBERT RECORDE.
THE EMINENT MATHEMATICIAN,
WHO WAS BORN AT TENBY, circa 1510.
TO HIS GENIUS WE OWE THE EARLIEST
IMPORTANT ENGLISH TREATISES ON
ALGEBRA, ARITHMETIC, ASTRONOMY, and GEOMETRY:
HE ALSO INVENTED THE SIGN OF
EQUALITY = NOW UNIVERSALLY ADOPTED
BY THE CIVILIZED WORLD.
ROBERT RECORDE
WAS COURT PHYSICIAN TO
KING EDWARD VI. and QUEEN MARY.
HE DIED IN LONDON.
1558.

SYMBOL	DATE	SOURCE
+ (plus) – (minus)	1489	Johannes Widmann, Germany, *Rechnung auf allen Kauffmanschaften.*
√ (square root)	1525	Christoff Rudolff, Germany, *Die Coss.*
= (equals)	1557	Robert Recorde, England, *The Whetstone of Witte.*
× (multiply)	1618	William Oughtred, England, in an appendix to Edward Wright's translation of John Napier's *Descriptio.*
a, b, c for known quantities (constants) *x, y, z* for unknown quantities (variables)	1637	René Descartes, France, *Discours de la méthode pour bien conduire sa raison et chercher la vérité dans les sciences.*
÷ (divide)	1659	Johann Rahn (or Rhonius), Germany, *Teutsche Algebra.*

> *"I will sette as I doe often in woorke use, a paire of paralleles, or Gemowe lines of one lengthe, thus: ==, bicause noe. 2. thynges, can be moare equalle."*
>
> Robert Recorde

THE BEGINNING OF MODERN MATH?

For all the importance of Stifel's *Arithmetica integra*, it was to be superseded within the year. In 1545 a work appeared that was so revolutionary in its central concept that some people have taken it to mark the start of the modern period in mathematics. In *Ars magna*, Gerolamo Cardano (see panel opposite) explained how to solve cubic (third-order) and even quartic (fourth-order) equations. However, it was not a straightforward triumph of individual genius. The solution to cubics had probably been discovered by Scipione del Ferro (*ca*.1465–1526) a professor of mathematics at the University of Bologna. On his death, he passed the information to a student, Antonio Maria Fior. Niccolo Tartaglia (*ca*.1500–57) independently discovered the solution and disclosed it to Cardano on condition that he did not reveal it. When Cardano discovered the solution was not original to Tartaglia, he published it. He did, admittedly, own that he had a clue from Tartaglia. He also acknowledged that the solution of the quartic had been by his amanuensis, Ludovico Ferrari (1522–65). Tartaglia, as can be imagined, was not well pleased and the two battled for ten years. Tartaglia had hoped to retain the revelation of the cubic to publish as the crowning achievement of his career. (Tartaglia had, previously, published the findings of others without acknowledging his own debt, which may reduce our sympathy for him a little.)

Cardano was a little more open-minded with regard to roots of negative numbers than most of his predecessors. Although he just about entertained the possibility of a negative root, he dismissed it as being "as subtle as it is useless."

Cardano's book represented the greatest advance in algebra since the Babylonians had discovered how to solve quadratic equations by completing the square. Although it was of no practical use—indeed, the solution of cubics by successive approximation by Jamshid al-Kashi (1380–1429) was more useful than Cardano's method—it stimulated further development of algebra and took the subject beyond the realm of the physical world. If quartics could be solved, then why not fifth-order equations, sixth-order

It seems likely that Gerolamo Cardano benefited from the work and ideas of other mathematicians in his groundbreaking book Ars magna (1545).

GEROLAMO CARDANO (1501–76)

Born in Pavia, Italy, Gerolamo Cardano was the illegitimate child of Fazio Cardano, a friend of Leonardo da Vinci. His mother tried to abort him, and his three siblings died of plague. After some difficulties in being accepted, he trained as a doctor and was the first to describe typhoid. He became professor of medicine at Pavia in 1543 and at Bologna in 1562.

As well as being a physician, Cardano was one of the foremost mathematicians of his day. His publication of solutions to cubic and quartic equations in *Ars magna* secured his place in history, but he also published the first systematic work on probability a hundred years before Pascal and Fermat.

Cardano's private life was colorful, and certainly fed into his interest in probability. He was always short of money and supplemented his income by gambling and playing chess. His treatment of probability, which he applied to gaming, includes a section on how to cheat effectively.

Life was not easy for Cardano. His favorite son was executed in 1560 for

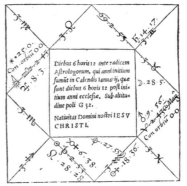

Cardano's horoscope of Jesus Christ that got him into so much trouble. A comet in the ascendant Libra can be seen as an interpretation of the star of Bethlehem, while the star Castor in Gemini predicted violence within Christ's life.

poisoning his wife. In 1570, Cardano was accused of heresy and imprisoned for several months for calculating the horoscope of Jesus Christ. As a consequence he lost his professorship at Bologna, as well as the right to publish books. He died on the day he had previously predicted, although he might have aided the fulfillment of his prophecy by committing suicide.

equations, and even higher? Suddenly, algebraic problems no longer needed to relate to real-world problems in the dimensions we recognize. Further dimensions, for the sake of mathematics, could be postulated, at least in theory.

While further dimensions were clearly absurd to Cardano's contemporaries, of interest only in the arena of fantastic mathematical exploration, they would come into their own several centuries later. By opening up the possibility of algebra and algebraic geometry extending into more than three dimensions, Cardano laid the foundations for Riemann geometries and the four-dimensional space-time continuum with which Einstein would remodel the universe (see pages 115–9).

Algebra Comes into its Own

The golden age of European algebra, which began with Cardano's publication of the solution of cubics and quartics, encompassed the legitimization of negative and complex numbers, the development of the Cartesian coordinate system, the marriage of algebra and geometry in analytic geometry, as well as considerable steps toward the development of integral calculus.

British mathematicians came into their own again after a long absence from the scene, but did not displace the Italian, German, and Polish mathematicians. Some of these men were now writing in their own languages rather than Latin.

Toward Complex Numbers

Soon after the Cardano-Tartaglia solution of cubics and quartics appeared, the Italian mathematician Rafael Bombelli ($ca.1526-72$) became the first to calculate using complex numbers. (Complex numbers are those that involve the square root of -1, i.)

Working with cube roots, he developed equations that used imaginary roots as a stage in deriving final solutions that are real numbers. He described it as "a wild thought" and it did not in fact help in his computations, but it did signal the importance that complex numbers were to have for algebra in the future.

Dealing with Numbers and Notation

Despite all their advances, the algebraists and trigonometers of the 16th century still did not have a widely used notation for decimal fractions. When Rheticus began his most ambitious trigonometric tables, he used triangles with sides of length 10^{15} units to attain the degree of accuracy he wanted without having to use fractions of any kind. (It doesn't matter which units as he didn't actually construct the triangles, just suggested them.)

François Viète (see box opposite) was only a part-time mathematician, but made progress in various fields—arithmetic, trigonometry, geometry, and, most importantly, algebra. He was instrumental in bringing about changes in notation that made further progress possible and promoted the use of decimal rather than sexagesimal fractions. Viète's most important contribution was in bringing consistent notation to algebra. This enabled him to develop a systematic way of thinking and a new method of working with general forms of equations. He adopted vowels to represent unknown quantities and consonants to represent known quantities. He also showed how to change the form of equations by multiplying or dividing each side by the same magnitude. For example, he showed how to transform the equation

$$x^3 + ax^2 = b^2x$$

into

$$x^2 + ax = b^2.$$

Viète still did not recognize negative or zero terms, so he could not reduce the number of possible equations to a single form in each order. (We have the form $ax^2 + bx + c = 0$ as the standard form which can describe any quadratic equation because, by allowing a,

FRANÇOIS VIÈTE (1540–1603)

François Viète was a French mathematician and Huguenot sympathizer. Trained in law, he became a member of the Breton parliament, then of the King's Council serving Henri III and Henri IV. He was proficient at deciphering secret messages intercepted by the French. Indeed, he was so successful that the Spanish accused him of being in league with the devil, complaining to the Pope that the French were using black magic to help them win the war.

Viète made great advances in several fields of mathematics, but always working in his spare time. Being wealthy, he printed numerous of his papers at his own expense. For a period of nearly six years in the second half of the 1580s, he was out of favor at court and concentrated almost exclusively on mathematics. In the 12th century, al-Tusi had found the same method of approximating roots of equations as that discovered by Viète.

b, or c to be negative or zero, it covers such possibilities as $x^2 - 7 = 0$, where b is 0 and c is negative.)

It is impossible to overstate the importance of good, consistent notation for the progress of algebra. Yet this was not Viète's only achievement. He arrived at formulae for multiple angles, was the first person to use the law of tangents (although he did not publish it) and the first to see that trigonometry could be used to solve cubic equations that could not be reduced. He also produced the first theoretical precise numerical expression for π:

$$\frac{2}{\pi} = \sqrt{\frac{1}{2}} \times \sqrt{\frac{1}{2} + \frac{1}{2}\sqrt{\frac{1}{2}}} \times \sqrt{\frac{1}{2} + \frac{1}{2}\sqrt{\frac{1}{2} + \frac{1}{2}\sqrt{\frac{1}{2}}}} \cdots$$

Although the method is not new, it was the first time the infinite series had been expressed analytically. Algebra and trigonometry were moving more and more

toward a concern with the infinite—both the infinitely large and the infinitely small.

Progress accelerated as a clutch of talented mathematicians applied themselves to developing algebra in its new directions. French mathematician Albert Girard recognized that the number of roots an equation has depends on the order of the equation—so a second-order equation has two roots, a third-order equation has three roots, and so on. The breakthrough came because he was sufficiently open-minded to allow negative and imaginary numbers in roots. Englishman Thomas Harriot (1560–1621) introduced the symbols > and < for greater than and less than. He was also the first proper mathematician to set foot on American soil, having been sent in 1585 by Sir Walter Raleigh as a surveyor. More influential than Viète in promoting the adoption of decimal fractions was the Flemish mathematician Simon Stevin (1548–1620). He also urged the adoption of a decimal system of weights and measures, though this was not to happen for another 200 years. Stevin adopted a notation for powers similar to that in use now, using a number in a circle raised above the line to show the power—so $5^{②}$ means 5^2. He even used fractional powers to show roots, so $5^{1/2}$ means $\sqrt{5}$. But Stevin was primarily a practical mathematician, and he dismissed any consideration of complex numbers.

The confidence with which the best mathematicians now approached algebra—and the distance it had traveled from its roots in real-world geometric problems in up to three dimensions—is clear in the public challenge set in 1593 by the Belgian mathematician Adriaen van Roomen (1561–1615) to solve a 45th-order equation:

$$x^{45} - 45x^{43} + 945x^{41} - \ldots - 3795x^3 + 45x = K$$

No concept of 45 dimensions was needed. Viète rose to the challenge and solved the equation when an ambassador to the court of Henri IV said that there was no Frenchman capable of it.

THE APPROACH TO ALGEBRAIC GEOMETRY

Viète's solution related to sines and he used his multiple-angle formulae to derive it. In providing a consistent symbolic system for representing algebraic equations, he also made it a matter of choice whether one solved a problem by geometric or algebraic methods. By bringing trigonometry to bear on algebra he was widening the scope of the subject and promoting its alliance with geometry. Viète was in fact one of the first people to view mathematics as a unified whole rather than different branches to be considered separately.

In 1572, Bombelli's *Algebra* had presented many geometric problems which he solved algebraically. For example, he gave algebraic solutions of cubics and then showed geometric demonstrations of his solutions. (However, this part of his treatise was not included in the printed edition and didn't appear until 1929.) Seventy-five years later, Descartes would take geometric problems, convert them to an algebraic form to simplify them as far as possible, then return to geometry for a final solution. In this, his analytic geometry, he completed a journey begun by Apollonius when he showed that conic sections could represent quadratic equations.

> "There are enough legitimate things to work on without the need to get busy on uncertain matter."
>
> Simon Stevin, 1585

L'ALGEBRA
OPERA
Di RAFAEL BOMBELLI da Bologna
Diuisa in tre Libri.
Con la quale ciascuno da se potrà venire in perfetta
cognitione della teorica dell'Arimetica.
Con vna Tauola copiosa delle materie, che
in essa si contengono .
Posta hora in luce à beneficio delli studiosi di
detta professione .

IN BOLOGNA,
Per Giouanni Rossi. MDLXXIX.
Con licenza de' Superiori ,

The title page to a 1579 edition of Bombelli's Algebra. *The first three volumes of an intended five were published in 1572. Bombelli died that year before he could finalize the last two volumes.*

GIANTS OF THE 17TH CENTURY

From the first half of the 17th century there was more communication between mathematicians than there had been at any time since Plato's Academy. In many countries, mathematical societies grew up alongside the other learned societies then appearing. In Britain, the mathematical society had the enticing name of the "Invisible College." In France, communication was further facilitated by Father Marin Mersenne, who corresponded with hundreds of mathematicians, scientists, and other learned men, acting as a conduit for knowledge and a sort of early networking guru. This meant that there were fewer incidences of mathematicians privately developing work that was then lost and had no impact on others. Mersenne facilitated disagreement as much as anything, but at least no one was in any doubt about what everyone else was doing. By a process of steady accretion, the foundations of modern mathematics were laid. Two men, both French, were to play a leading role in that process.

Neither of the two towering figures of the age was a professional mathematician. René Descartes (see page 138) was a minor scion of the French nobility who is more famous as a philosopher than as a mathematician. His explanation of his system of analytic geometry is provided in an appendix to his philosophical text, *Discourse on Method*, as a demonstration of

Marin Mersenne, who saw it as his Christian duty to disseminate scientific knowledge.

how he used reason to arrive at his results. Pierre de Fermat (see page 139) was a lawyer and then a councilor who pursued his interest in mathematics in his spare time. Yet his ability rivaled that of Descartes.

MARRYING ALGEBRA AND GEOMETRY

Descartes found neither geometry nor algebra entirely satisfactory and set about taking the best of both. By seeing the quantities in his equations as line segments, Descartes avoided any conceptual difficulty in working with higher-order equations and dealing with equations that did not have expressions of the same order on each side. For example, the Greeks could not allow an equation such as $x^2 + bx = a$ because the two parts on the left-hand side are considered areas and that on the right is considered a line; an area and a line cannot be considered equal.

Descartes refined Viète's notation, using letters near the start of the alphabet for known quantities (a, b, c) and letters near the end of the alphabet for unknowns (x, y, z). He used raised numbers to indicate powers and used the symbols for the arithmetical operators that we still use. Only his symbol for equality was different as he had not adopted Robert Recorde's pair of parallel lines.

137

RENÉ DESCARTES (1596–1650)

René Descartes was born in Tourraine, France. His mother died when he was only a year old. His father remarried and moved away, leaving the infant Descartes in the care of relatives. He trained in law, taking his degree in 1616, and then traveled. It was while he was in Bohemia in 1619 that he developed analytic geometry.

Descartes shared some views and practices of the mystical group the Rosicrucians. Like their full followers, he moved around a good deal, always lived alone, and practiced medicine without charge, but he rejected their mystical beliefs. He promoted religious tolerance and championed the use of reason in his scientific and philosophical writings.

Descartes has been called the father of modern philosophy for his contention, propounded in his *Discourse on Method,* that knowledge must be acquired through reasoning. He maintained that sensory perceptions are not a reliable guide to the world around us and cannot be depended upon to yield true information. His famous dictum, "I think, therefore I am," is part of his demonstration of the few things that can be relied upon—the existence of the thinking mind, of God, and of the material world. The dichotomy between mind and body was another of his preoccupations. His belief in free will was paramount; he adopted the anti-Calvinist view that salvation can be earned through the operation of free will and does not depend only on God's grace.

Descartes was always sickly and, when Queen Christina of Sweden invited him to her court to teach her philosophy, and demanded that he get up at 5 AM each day, he quickly succumbed to the Scandinavian winter and died.

An engraving of Queen Christina of Sweden, whose unreasonable demands for instruction by the great philosopher precipitated Descartes' death.

PIERRE DE FERMAT (1601–65)

Born in the Basque region of France, Fermat studied law and later mathematics. He developed independently of Descartes the principles of using a coordinate system to define the positions of points.

Fermat worked extensively on curves, developing a method for measuring the area under a curve that is similar to integral calculus, and to generalized definitions of common parabolas. He worked extensively, too, on the theory of numbers and corresponded with Blaise Pascal on this subject. This was his only contact with other mathematicians. He was a secretive recluse, who generally communicated only with Marin Mersenne (see page 137).

Fermat was the most productive mathematician of his day, but was so reluctant to publish that he gained little credit for his work during his lifetime.

An engraving of Pierre de Fermat taken from Louis Figuier's Vies des Savants Illustres (Lives of the Great Scientists), *of 1870.*

Descartes proposed that the position of a point in a plane could be identified by reference to two intersecting axes, used as measuring guides, so developing the coordinate system that is now known as the Cartesian system. For all the familiarity of his algebraic notation, Descartes' graphical representations of equations do not all resemble ours, for he never used negative values of x in his graphs. The familiar form of a graph divided into quadrants by axes that cross at (0,0) was introduced later by Isaac Newton. In addition, his axes were not always set at right angles to each other.

Descartes believed that any polynomial expression in x and y could be expressed as a curve and studied using analytic geometry.

At the same time as Descartes was formulating his analytic geometry, another Frenchman, Pierre de Fermat, was doing much the same thing. Both arrived at comparable results independently. Fermat stressed that any relationship between x and y defined a curve. He recast Apollonius' work in algebraic terms, aiming to restore some of Apollonius' lost work. Both Descartes and Fermat proposed using a third axis to model three-dimensional

British mathematician Andrew Wiles, who proved Fermat's Last Theorem, is a professor at Princeton University. He received a knighthood in 2000.

of the semicubal parabola (a method for discovering the length of a curved line).

FERMAT'S LAST THEOREM

Fermat is most famous now for his so-called "last" or "great" theorem. He noted in the margin of his copy of Diophantus' *Arithmetica* that there are no solutions to the equation

$$x^n + y^n = z^n$$

for values of n greater than 2. He added, "I have discovered a truly marvelous proof of this, which, however, the margin is not large enough to contain"—and so the proof was lost and the subsequent search for it taxed mathematicians for more than 300 years. Because the problem is so easy to understand, many people tried to solve it before it was finally mastered by the English mathematician Andrew Wiles in 1995. Wiles proved Fermat's theorem with a method that uses elliptic curves. He had tried to solve it when he was still only a child, as soon as he heard about it, and continued through his degree course in mathematics. Long after he had given up he realized that it was related to his work on curves and returned to the problem again. His proof is highly complex and could not have been the same as that Fermat claimed to have found.

curves, but this was not advanced until later in the 17th century.

Neither Descartes nor Fermat sought to publicize their work widely. Descartes did publish his, writing in French so that more people could understand it, but he did not explain in great detail and much of the work was impenetrable to many readers. It is not entirely clear whether Descartes wanted to exclude people whom he felt weren't sufficiently serious or whether he wanted to give his readers the pleasure of discovery by making some of the intellectual leaps and bounds themselves, but either way it did little to help the dissemination of his ideas. Soon, an anonymous introduction was added to his work to help explain it. In 1649 Frans van Schooten published a Latin edition with explanatory commentary.

Fermat was little better at promoting his work than Descartes, being a confirmed recluse who refused to publish. Dissemination of his ideas during his lifetime was almost exclusively through the mediation of Marin Mersenne; indeed, Fermat published only one of his discoveries, the rather obscure rectification

The World Is Never Enough

Descartes brought algebra and geometry together by defining a point by coordinates and using this to draw graphs from equations. In doing this, he provided the means for a later development of algebraic geometry into untold new dimensions.

Any two-dimensional shape can be represented by giving the coordinates of its vertices (corners), each as two numbers. The principle can be extended to three dimensions easily—by giving three coordinates we define a point in three-dimensional space. It is easy to work out the differences between points, too. In a two-dimensional system, with points (a,b) and (c,d) we can use Pythagoras' theorem to work out the distance between the points. We imagine a triangle, with the two points defining the ends of the hypotenuse. The length of this line—the distance between the points—is then $\sqrt{((c—a)^2 + (d—b)^2)}$. We can extend the same formula to three dimensions: the distance between the points (a,b,c) and (d,e,f) is $\sqrt{((d—a)^2 + (e—b)^2 + (f—c)^2)}$. What is there to stop us taking this further and dealing with distances in four dimensions, defined by four coordinates? Or 26 dimensions? Or 4,519 dimensions? We may have a conceptual objection because we can't visualize four-dimensional space, still less 4,519-dimensional space, but mathematics is not concerned with whether we are comfortable with the concept.

What use is multidimensional space? If we can step back from the problems of trying to visualize it as a real-world space, the theoretical space with many dimensions is actually quite useful. We often draw graphs that plot two variables—speed against time, for example, or temperature against growth rate. There are many situations in the real world in which far more than two variables are involved. If we track weather conditions, or the performance of companies in a stock market, or the mortality rates in a population, there are many variables to take into account. By allocating values for perhaps seven, eight, or nine variables to each data point we can envisage, if not visualize, a map in seven, eight, or nine dimensions from which we can make measurements and predictions could be made. It isn't necessary to draw the map—algebra can take care of the calculations without that—but the conceptual space has been suggested in which the graph exists.

It is very hard for us to visualize space with more than the four dimensions we know to exist, but algebra can work in any number of dimensions.

THE KOCH SNOWFLAKE

It is even possible to conceive of geometry in fractional dimensions. A famous model of this is the Koch snowflake, developed by the Swedish mathematician Niels von Koch (1870–1924). The Koch snowflake is an example of a fractal, one of the earliest defined.

Draw an equilateral triangle; divide each side into three equal portions. Remove the middle portion from each side, replacing it with two sides of another equilateral triangle the same size as the removed section. Keep doing this. The result is a shape like a snowflake.

It's possible to carry on doing this an infinite number of times. The result is a shape that has an area defined by the formula

$$\frac{2\sqrt{3}s^2}{5}$$

when s is the measure of one side of the original triangle. However, the perimeter is infinite—an infinite perimeter encloses a finite area.

Carrying out the same operation with a single line segment instead of a triangle, the resulting line approaches a curve as the line segments get smaller and smaller. The curve is called a Koch curve.

The curve has infinite length. The total length increases by one third at each step and so the length after n steps is $(\frac{4}{3})^n$. It is not a one-dimensional line, as any portion is unmeasurable—it is infinitely long. Yet it is not enclosing an area, so it is not two-dimensional either. It is said to have a fractal dimension of $\log 4/\log 3 \approx 1.26$, greater than the dimension of a line, but less than the dimension of a curve. (A fractal dimension is also called a Hausdorff dimension after one of the founders of modern topology.)

OTHER FRACTALS

A fractal is a structure in which a pattern is repeated from the large scale to the small scale, so that looking more closely at the structure reveals the same or similar figures. There are many near fractals in nature, including snowflakes, trees, galaxies, and blood-vessel networks. Fractals are too irregular to be described using standard Euclidean geometry and generally have a Hausdorff dimension that differs from their normal topological dimension. Fractals are often produced by space-filling algorithms. The Sierpinski triangle is an example. Starting with a simple triangle, make three copies of it at one half the size of the original, and place the copies in the corners of the original. Carry on repeating this step *ad infinitum*. The resulting pattern is identical at any magnification. It was first described by the Polish mathematician Waclaw Sierpinski (1882–1969) in 1915 in the form of a mathematically defined curve rather than a geometric shape. It has a Hausdorff dimension of log 3/log 2 ≈ 1.585.

> *"Clouds are not spheres, mountains are not cones, coastlines are not circles, and bark is not smooth, nor does lightning travel in a straight line."*
>
> Benoît Mandelbrot

The best known example of a fractal is the Mandelbrot set, described by the Polish mathematician Benoît Mandelbrot (1924–2010). This is the result of drawing a geometric figure of a set of quadratic equations that involve complex numbers.

Mandelbrot drew together earlier examples of fractals, gave them the name "fractal," and defined their conditions. He explored their prevalence, both in the natural world and in artificial systems such as economics, and determined that they are a very common model, more frequently found than the simple structures of Euclidean geometry. Fractals can often express the "rough" quality of the real universe, whereas Euclidean geometry deals with smoothness, which is rarely found in nature. Mandelbrot suggested a model of the universe in which stars are fractally distributed. This would solve Olbers' paradox without the need for a Big Bang, though it does not preclude a Big Bang. (Olbers' paradox states that the night sky is dark when it should be bright, since looking in any direction we should see a star. Although it was described by the German astronomer Heinrich Olbers in 1823, it was first noted by Kepler.)

Although fractals generally begin as equations, they are best realized as geometric shapes.

MOVING ON

With fractals, lines expand into infinity. Working with the graphs produced by Descartes and Fermat, even without the added complexity of infinite line lengths, soon produced a need to calculate the areas under curves and the lengths of curved line segments. The way of dealing with this—and, later, of dealing with fractals—involved looking to the infinitely small. In the late 17th century, mathematicians finally came to grips with the idea of infinity.

Benoît Mandelbrot, "the father of fractal geometry."

CHAPTER 6

Grasping the
INFINITE

Geometric methods for finding areas and
volumes are easy enough when dealing with
polygons and solids with straight edges, but
they fall down when confronted with curved
areas and volumes as well as with the even
more challenging spaces and surfaces of
Riemann geometry and fractals.

   Early approaches to the problems of
working with irregular shapes and volumes
tried to divide the area or volume into small,
regular parts then add together the parts.
The essential elements of this method were
described by Eudoxus and Archimedes more
than 2,000 years ago, but rigorous development
and application were not possible while
mathematicians still baulked at the idea of
infinity. The late 17th century finally saw a
systematic method for dealing with these
problems. As happened with analytic geometry,
the new method—calculus—was developed
simultaneously and independently by two of
the greatest mathematicians of the time.

The ceaselessness of the sea is an earthly symbol of the infinite.

Coming to Terms with Infinity

Irrational numbers, like π, e, and √2, are infinite series. We can go on refining them to ever more decimal places, but the task will never be completed. Both the infinitely large and the infinitely small (the infinitesimal) had worried mathematicians for two millennia. The Greeks disliked irrational numbers to the point, perhaps, of murdering Hippasus for proving their existence. But in the 17th century mathematicians made moves to approach and eventually embrace the infinite and the infinitesimal. These concepts and numbers were finally to become useful rather than just confounding to expectations and beliefs that were held dear.

Archimedes, pictured here in an anachronistic portrait of 1620, confronted problems of infinity and limits that would be addressed nearly 2,000 years later.

AN EARLY PRECURSOR

The method Archimedes adopted for calculating the area of a circle (and so obtaining a value for π) depended on drawing polygons inside and outside a circle and calculating their respective areas. These gave upper and lower limits for the area of the circle. A greater degree of accuracy was achieved by using ever-larger numbers of sides for the bounding and inscribed polygons. Here Archimedes encountered two concepts that would become hugely important later—that of limits and that of infinity, for the perfect area would be given by a polygon with infinitely many sides. A circle may indeed be called a polygon with infinite sides, so the two polygons would converge at that point. As the number of sides tends toward infinity, the difference between the area of the polygons and the area of the circle tends toward zero and the limits coincide.

THE WEIGHT OF PAPER

Just as Archimedes had found the volume of an irregular shape by measuring the volume of water it displaced, so Galileo discovered a practical solution to the problem of finding the area under a curve. In the absence of geometric and algebraic tools to calculate the area, he would plot his curve, then cut it out and weigh the paper. By comparing the weight with the weight of a piece of paper of known area, he could work out the area of his curve.

The possibility of working out an area or volume by dividing a figure into a very large number of very thin slices was not new to Archimedes. Democritus had rejected it 200 years earlier as he could not work around his objection to the logical difficulty that, if the slices are infinitely thin, there is no difference between them, so every pyramid becomes a cube. Antiphon developed the technique into the "method of exhaustion" (though that term was not used until 1647) and Eudoxus made it rigorous. The principle was to relate the area to be discovered to another area, which was easier to calculate, and prove first that the unknown area is not greater than the known area and then that it is not smaller than the known area (so they are equal). It is a nonconstructive method of proof, since the answer must be known before the proof can be used.

In the 17th century, when mathematicians finally became more comfortable thinking about the infinite and the infinitesimal, the method finally came into its own with proper algebraic formulation and emerged as integral calculus. This could not happen until analytic geometry had been developed and a rigorous understanding of limits had emerged.

STEPS IN THE RIGHT DIRECTION

In the second half of the 16th century, the great rush of development in science and mechanics brought new incentives for calculation with areas, volumes, and properties such as velocity. The German scientist Johannes Kepler (1571–1630) and the Flemish engineer Simon Stevin (see page 28) both worked on calculating the areas of irregular shapes by dividing them into very thin slices and both approached the problem from a practical point of view with a specific problem in mind. Stevin used the technique to address the problem of calculating the center of gravity

A 19th-century engraving of Democritus with straight edge, compass, and globe.

of a solid object. He inscribed parallelograms inside a triangle to find the median on which the center of gravity would lie.

Kepler had an altogether more interesting question. When paying for wine by the barrel, the price was calculated according to how full the barrel was. But this was measured with a dipstick and took no account of the vertical curvature of the barrel. Only at the point when the barrel was exactly full or half-full did the dipstick give an accurate measure of the volume, since the barrel is wider in the middle than at the top or bottom. If a barrel was a quarter full (in depth), it contained less than a quarter of a full barrel and Kepler would be cheated if he paid a quarter the price of a full barrel. He proposed cutting the barrel into infinitely thin circular slices and adding

up their areas as a method of calculating the true volume. In fact, he also needed to measure the areas under curved paths for his work on astronomy—but the wine barrel presents a more compelling problem.

Galileo stated his intention of writing a treatise on the infinite, but if he ever did so it has not survived. Instead there are passages that relate to calculating areas and volumes by reference to infinity and infinitesimals, but Galileo still grappled with the strange logic of these concepts.

Perhaps his most interesting observation, and one that looked forward to the development of set theory in the 19th century, was that each integer can be squared, so, since there is an infinite number of integers, there is an infinite number of squares: "We must say that there are as many squares as there are numbers." But can the infinite number of squares be larger than the infinite number of integers? He had come close to recognizing a feature of infinite sets—that a part of the set can be equal to the whole set. However, he backed away from this conclusion, saying instead that "the attributes 'equal,' 'greater,' and 'less' are not applicable to infinite, but only to finite quantities."

It was another Italian, Bonaventura Calvieri (1598–1647), who drew together the work on infinite division from Archimedes to Galileo. In a text published in 1635 (though conceived six years earlier), he explained his method of "indivisibles." He used laborious geometric methods that were soon superseded, but achieved an impressive result. He managed something that was equivalent to the calculus that would be invented only 50 years later.

> "[The infinite and infinitesimal] transcend our finite understanding, the former on account of their magnitude, the latter because of their smallness; imagine what they are when combined."
>
> Galileo, 1638

Calculus used to be called "the calculus" and today it is often used to solve complicated problems that algebra alone can not deal with.

The Emergence of Calculus

The invention of calculus was one of the great turning points in the history of mathematics. It tackled problems that had taxed mathematicians for 2,000 years and opened doors that no one even knew existed before.

A Bit about Calculus

Calculus provides a way of measuring rates of change and the effects of change. ("Calculus" is the Latin name for a small stone used for counting.) It is divided into two parts, which are the inverse of each other: differentiation and integration. The fundamental theorem of calculus is that applying differentiation to an integral returns the original expression, and vice versa.

Both are essentially methods of approximation, but aim to use limits that make the error involved (the inaccuracy of the approximation) tend toward zero. The principle is easier to understand when illustrated by an example.

Back to Achilles and the Tortoise

The paradox of Zeno, in which Achilles can never catch up with the tortoise if the tortoise is given a head start (see page 78), can be expressed (but not solved) using calculus.

Using d to represent the distance from the starting point that the tortoise has traveled, and t to represent the amount of time that has passed, we have a sequence of times and corresponding distances, t_1, t_2, t_3... and d_1, d_2, d_3.... The speed at which the tortoise moves is a function of time and distance, and gives the rate of change in the tortoise's position. His speed over an interval between two times, t_1 and t_2, is given by:

$$\frac{d_2 - d_1}{t_2 - t_1}$$

If after 15 seconds the tortoise is 3 feet from the starting block and after 20 seconds he is 4 feet from the starting block, his speed is

$$\frac{4 - 3}{20 - 15}$$

or ⅕ feet per second.

A graph showing the tortoise's movement would be a straight line, as the relationship between distance and time is constant.

If the tortoise moves at a steady speed, a graph of speed against time would be a straight horizontal line.

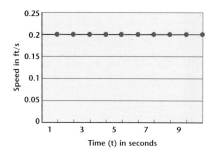

Time (t) in seconds

EARLY DYNAMICS

The French bishop Nicholas Oresme discovered *ca.*1361 that the area under a graph of speed against time is equal to the distance traveled. In his conversion of a problem in dynamics to geometry, he was probably the first to use a coordinate system outside cartography.

The distance covered is the area under the graph, speed vs. time, which is easy to calculate in this instance (d = 0.2t). The rate of change of speed (acceleration) is given by the slope. In this case, the line is flat, as there is no acceleration—the tortoise goes at uniform speed.

Now assume the tortoise has been given an electric scooter. Instead of a uniform speed, he now accelerates until the scooter reaches its top speed. The first part of the speed graph looks like this:

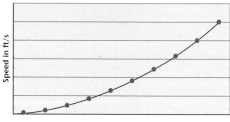

Time (t) in seconds

The situation is altogether more complex. To find the distance the tortoise has covered, we need the area under the graph, but this is not easy to calculate. To find the acceleration at any particular instant, we need to measure the slope of the curve at that point. The first is solved by integral calculus and the second by differential calculus.

INTEGRATION

Integration finds the area under the curve by drawing a series of infinitesimally thin rectangles under the curve and adding together their areas. It's very similar to Kepler's slices of wine barrel (see page 148) or the slices of pyramid that troubled Democritus (see page 79).

We can make a rough approximation of

the area under the curve by drawing rectangles so that the curve passes through the midpoint of the top of each rectangle:

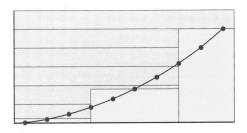

The line cuts off part of the top of each rectangle to the left, but there is a space under the line to the right. If the spare bit of rectangle were flipped over, it would fit the space pretty well. The smaller the rectangles, the better the fit to the curve.

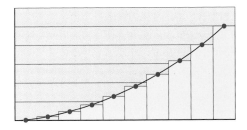

The calculated area (the sum of all the rectangles) approaches the true area under the curve as the number of rectangles approaches infinity. This area is the integral of the function f (t). The expression for the integral is written as

$$\int_a^b f(t)dt$$

where a and b are the limits we are working within (the upper and lower values of t that bound the area) and "dt" means a very small change in time.

DIFFERENTIATION

The average acceleration over an interval of time (on the graph of time against speed) is given by the slope of a straight line drawn between the start and end points of the interval. This line is called a secant. The acceleration at an instant is given by the slope of the curve at that instant (or of a tangent to the curve).

Differential calculus provides a way of approximating the slope of the curve by assuming a very short time interval and calculating the slope of the secant for that interval. (The very short interval is called Δt, "delta-t," the Greek capital delta being used to show a small quantity.)

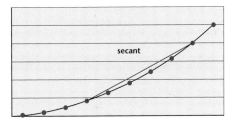

In this example, the secant is drawn between two points on the line.

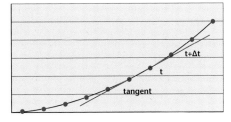

In this example, the tangent touches the line at a point.

The time Δt is a very short interval. Making Δt smaller and smaller produces a more and more accurate result, though it will never be quite the same as the slope of the curve because we can't set Δt to zero. However, as Δt approaches zero, the line approaches a perfect match. This introduces the concept of the limit: the limit of the function *approaches* the required value (the acceleration at an instant) as Δt approaches zero. This is the process of differentiation.

NEAR MISSES

Fermat's work on analytic geometry includes a relationship that is fundamental to the theory of calculus. Fermat dealt with finding tangents to curves and areas under curves. The expressions he derived had an inverse relationship, yet it seems to have escaped his attention, for there is no evidence that he pursued it or tried to explain it.

Blaise Pascal is another who could easily have taken a final step and discovered calculus. Pascal's interests in mathematics were varied and he flitted from topic to topic. He also gave up mathematics after he underwent a religious ecstasy, and he died young—two further factors that contributed to rob him of the prize that might have been his. Pascal came so close to discovering calculus while working on an integration of the sine function that Leibniz later wrote that it was reading Pascal's work that signposted calculus for him.

SEEING THE WAY FORWARD

With the development of analytic geometry, it became possible to describe movement algebraically. The Ancient Greeks had introduced the idea of a curve as the path (locus) of a moving point. Algebraic geometry provided a tool for describing that locus in the form of an equation, generalizing about the shapes of the curves produced by different types of motion and identifying patterns that had predictive value. For example, Calvieri noticed that the area under the parabola defined by $y = x^2$, between 0 to a on the x-axis is $a^3/3$. Similarly, for the curve $y = x^3$, the corresponding area is $a^4/4$. It was not then difficult to guess that the general formula for the area under a curve $y = x^n$ is $a^{n+1}/(n + 1)$.

LEIBNIZ AND NEWTON

The fundamentals of calculus—both differentiation and integration—were discovered around 1670 independently by

---

SEKI KOWA OR SEKI TAKAKAZU (1637/42–1708)

Seki Kowa was born in Japan in either 1637 or 1642. He developed a new notation for expressing equations up to the fifth degree, using *kanji* characters for variables and unknowns. He discovered discriminants, which led him to some results in differential calculus at around the same time as Newton and Leibniz discovered them in Europe. There is no known communication between the European and Japanese mathematicians. (A discriminant is an expression that shows a relationship between the coefficients of a polynomial equation. For example, for the quadratic $ax^2+ bx + c$, the determinant is b^2—4ac. Whether the discriminant is positive, negative, or zero gives information about the nature of the equation and its roots.)

SIR ISAAC NEWTON (1642–1727)

Isaac Newton was born prematurely on Christmas Day in England in 1642 and was so sickly that he was administered the last rites. His father had died before his birth and, when he was three, his mother left him to be cared for by his grandmother while she went to live with her new husband.

Newton went to Trinity College, Cambridge, where he studied the classical science required by the curriculum but also read the new works of Descartes and the chemist Robert Boyle. When the university was closed for two years because of plague, Newton worked in Lincolnshire. He developed his ideas of calculus, which he called "fluxions," but at this point he did not publish any of his findings. After the plague, Newton became a professor at Cambridge and could

Newton was the first to realize that white light could be split into the colors of the spectrum.

dedicate himself to his scientific and mathematical work. He discovered that white light is made up of a spectrum of colored light, formulated his laws of motion (which underpin classical mechanics), and defined a force that directs the motion of falling bodies, gravity. His publication of his discoveries, *Philosophiae Naturalis Principia Mathematica,* is perhaps the single most important scientific publication of all time.

Newton was attracted to mystical matters and alchemy. He also had a psychotic intolerance of other scholars disagreeing with him. For many years, he worked in isolation, shunning any possible source of conflict. Professional disputes sometimes prompted him to burst into tantrums that took the scientific world by storm.

the English scientist and mathematician Isaac Newton and the German polymath Gottfried Leibniz (see page 154).

What both men did was to discover a method for calculating the tangent of a curve at a specified point on the curve, given only the equation defining the curve. The slope of the tangent (which defines the line geometrically) shows the rate of change of the function (such as the speed of a moving body at a particular instant). Both men also realized that integration is the reverse of this process of differentiation—that integrating the result of differentiation leads back to the original function and vice versa. This revealed a surprising relationship between total values and rates of change.

The notation

$$\int x^2 dx$$

for the expression giving the area under the curve y = x² was adopted by Leibniz because he saw it as the sum (indicated by the elongated s, ∫) of the expression, in this case x², divided into infinitely small segments along the x-axis (dx). Newton and Leibniz stressed different aspects of calculus and had quite separate intentions in using it. For Newton,

one of the benefits of his discoveries in calculus was the ability to tackle power series—infinite sums of multiple powers of x, such as

$$^1/(1-x) = 1 + x + x^2 + x^3 + x^4 + \cdots$$

He developed a calculus of power series, showing how to differentiate, integrate, and invert them. Leibniz was more interested in the properties of changing systems and in summing infinitesimals. His work treated continuous quantities as though they were

GOTTFRIED WILHELM LEIBNIZ (1646–1716)

Gottfried Leibniz was largely self taught as a child, then entered the university of Leipzig, Germany, to study law. The university refused him a doctorate because he was too young and he left the city, never to return. He was awarded his doctorate immediately at Nürnberg.

Leibniz moved to Paris, and most of his writings are in French or Latin. He worked in the service of several noble families during his lifetime, pursuing his interests in mathematics, philosophy, and many branches of science at the same time. He developed a calculating machine, which he presented to the Royal Society on a visit to London. He developed the

Leibniz wrote tens of thousands of documents on a wide range of subjects and much of his work remains unpublished to this day.

branch of science called dynamics, which is concerned with the movement of objects and the forces acting upon them, then worked in the 1670s in practical mechanics and engineering, designing and improving many kinds of machinery. He is considered the originator of geology after the observations he made at the mines in the Harz mountains. He was the first to propose that the Earth had initially been molten. In 1679 he perfected binary notation, which is at the heart of computer science. His philosophy was optimistic—he believed that the world represented the best of all possible worlds that God could have created.

Zeno's paradoxes center around dividing up continuous movement into tiny moments—a problem glossed over by calculus.

discrete, a logical flaw that he and others overlooked—even though it was an issue so old that it was the difficulty at the heart of Zeno's paradoxes.

Newton failed to publish any of his findings relating to calculus, and Leibniz published first. The ensuing dispute over priority and the merits of each man's methods had long-lasting repercussions, isolating British math until the 19th century.

THE IMPACT OF CALCULUS

In studying falling bodies, Galileo (who died the year Newton was born) needed to calculate the speed of an object at a particular instant in time. For this type of problem, differential calculus is the perfect tool. Since the time of Newton and Leibniz, calculus has been applied to countless problems in mechanics, physics, astronomy, economics, social science, and many other fields, revolutionizing physics and giving new impetus to the further development of mathematical techniques. Calculus has spawned a whole branch of mathematics, called analysis, which deals with continuous change. In summing a large set of small quantities, integration is useful in problems such as determining the distance traveled by a body moving at varying speed or calculating the total fuel consumption of a vehicle. Differentiation can be used in such varied problems as modeling disease epidemics and determining the path an aircraft needs to take to avoid colliding with another.

Calculus has wide-ranging practical applications for the modern world. It even helps to establish safe flight paths for aircraft in our busy skies.

Calculus and Beyond

Mathematicians had never been comfortable with the concept of infinity and calculus highlighted this concern. The Anglican bishop George Berkeley (1685–1753) made a well-argued refutation of calculus and this prompted a productive debate, which led to the rigorous definition of limits and infinity. This ultimately benefited the development of calculus and enabled analysis to grow out of it.

Berkeley's objection was not fully answered for more than a hundred years. A further century later, the logician Abraham Robinson (1918–74) finally showed that the idea of the infinitesimal is logically consistent and that infinitesimals can be considered a kind of number.

Analysis deals with continuous change and with processes that have emerged from studying it, such as limits, differentiation, and integration. In particular, differentiation is one of the principal tools of analysis. Relating rates of change to present values, it is possible—at least in theory—to predict future behaviors. This puts analysis at the heart of many modeling and predictive activities, from weather forecasting to epidemiology, from astronomy to fluid mechanics.

USING INFINITY

As long ago as the time of Ancient Greece, infinity had been employed in mathematics, despite the difficulty of such a concept. Euclid used an algorithm to find the greatest common divisor of a pair of numbers. If he applied it to a pair of

Bishop Berkeley responded to what he saw as the undermining of religion by natural philosophy (of which calculus was a key part) in The Analyst *(1734).*

irrational line segments, the procedure never terminated, becoming an infinite process. Euclid used this property to test irrationality.

A similar willingness to ignore the logical difficulties of infinity and infinitesimals (or lack of rigor in applying such logic) led Newton and Leibniz to fudge calculus. They treated reality as both discrete and continuous at the same time, depending on quantities that were so tiny they could wink out of existence when convenient, or could be used to put a stop to

> "And what are these Fluxions?... They are neither finite quantities, nor quantities infinitely small, nor yet nothing. May we not call them the ghosts of departed quantities?"
>
> George Berkeley, 1734

MADHAVA OF SANGAMAGRAMA (1350–1425)

The Indian mathematician Madhava of Sangamagrama is considered by many to be the earliest originator of analysis as a method. He founded the Kerala school of mathematics and astronomy, which flourished between the 14th and 16th centuries. He was the first to accept limits tending to infinity and to define infinite series. He discovered the infinite series of the trigonometric functions and developed several methods for calculating the circumference of a circle and two methods for finding π. He also made steps toward the development of both integral and differential calculus. His writings have not survived, so his achievements are known by reputation only.

infinity. However, the results achieved by calculus, in so many fields, were so valuable and impressive that the inconsistencies at the heart of calculus were not immediately addressed.

AFTER NEWTON AND LEIBNIZ

The disagreement between Newton and Leibniz over priority had the result that subsequent work with calculus was polarized. As Leibniz had come up with the usable notation, his was carried forward and Continental Europe became the arena in which developments in calculus were played out. Newton's interest in geometry obscured the calculus he was using and his aversion to his rival's notation meant that few people in Britain followed him, despite admiring the results he achieved.

In the decades after Leibniz's publication, the Swiss brothers Jacob (1654–1705) and Johann Bernoulli (1667–1748) dominated calculus, along with Leibniz himself. The Bernoullis developed the rules for differentiation, the integration of rational functions, the theory of

Jacob (left) and Johann (right) Bernoulli came from a talented family of tradesmen and scholars and were only two of eight mathematicians in the family!

elementary functions, applications to mechanics, and the geometry of curves—in fact, most of the fundamentals of classical calculus with the exception of the power series that most interested Newton. The Bernoulli brothers even used calculus to demonstrate Newton's own inverse square rule applied to gravity in an elliptical orbit, which Newton had not explained well.

In the mid-19th century, Riemann refined the method for calculating an integral, suggesting comparing two sets of thin slices, one inscribed and one circumscribed. As the two values approach each other (with thinner and thinner slices), the true integral is found.

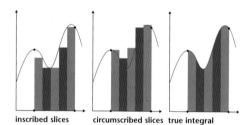

inscribed slices   circumscribed slices true integral

DEALING WITH DILEMMAS

However useful calculus was, the inconsistencies at its heart would not go away and sooner or later they had to be dealt with. Dealing with these resulted in the development of analysis—not itself a computational technique, but a sound logical basis for using calculus.

Two dilemmas, highlighted by the critics of early applications of calculus, arise as soon as we start to think more rigorously about differentiation than did Newton and Leibniz. One is evocatively captured in Berkeley's "ghosts of departed quantities"; the other might be called the "ghost of a moment." The real world is better

characterized by the model of a continuum than by a set of discrete parts. (This recalls the distinction we made at the very start between counting and measuring, between arithmetic and geometry, as well as the problems of the continuous and the discrete at the heart of Zeno's paradoxes.)

Think about any system that involves continuous change—water flowing over a dam or air over an aircraft wing, for example. As local conditions vary, the rate of flow is not constant. Measuring it at any moment involves some kind of approximation or averaging as the time interval could always be made smaller. Only by freezing time could we take an accurate measurement. But flow depends on time, so if we freeze time the flow is zero.

It is not only time that can be endlessly subdivided. For example, as temperature changes from 2° to 3°, it must go through an infinite number of intermediate stages; even 2° and 3° themselves are infinite decimals, with an infinite number of zeroes after the decimal point. In modeling continuous change, we must deal with these fleeting values—and they are necessarily infinite decimals.

The concepts and deductive structures behind infinite quantities came to preoccupy mathematicians working with calculus as they struggled to develop rigor.

For analysis to become a rigorous and dependable tool, mathematicians first needed to find some way of dealing with the vagueness of these ghosts of quantities and moments. The German mathematician Karl Weierstrass (1815–97) provided a satisfactory definition of the limit of a series. He became known as the father of modern analysis for devising a test for the convergence of series and for his work on functions. Using the series

½ + ¼ + ⅛ + ...

Weierstrass would say that all we need to do is pick the level of error (or approximation) that is acceptable (ε) and then continue with the series until the sum (so far) differs from 1 by smaller than ε. At this point, we say that the series converges to the limit 1. This removes the need for nebulous infinitesimals and gives a real number that satisfies the requirements. Also, although the series approaches its limit, it does not have to reach it for Weierstrass's condition to be met. Now, the margin of approximation could be stated and the

> ### INFINITE SERIES
>
> An infinite series is a series with an infinite number of terms. For example,
>
> ½ + ¼ + ⅛ + ...
>
> is an infinite series, with each term being half of the last. The limit of this series—the number that would be reached if we could get to the end of the infinite number of terms—is 1. Because the series reaches a definite limit it is said to converge. Other series do not converge, such as
>
> 1 + 2 + 3 + ...
>
> This series diverges as it never settles to a limit. Some convergent series can be ambiguous:
>
> 0 + 1—1 + 1—1 + ...
>
> oscillates between 1 and 0.

German mathematician Karl Weierstrass was concerned with eliminating inconsistencies in calculus and defining the limit of a series.

degree of accuracy quantified. There was no need to worry about quantities that had to disappear from existence—analysis was put on to a logical footing.

CALCULUS BECOMES ALGEBRAIC

During the 18th century, calculus moved away from its geometric roots in the work of Newton and Leibniz and became increasingly algebraic. Geometric curves became less important and algebraic functions moved to center stage. Soon, complex numbers moved in on the scene.

Differentiation offers a useful tool for

finding local maximum and minimum values between upper and lower limits. If we draw a curve of a function, the slope approaching a maximum point flattens out; the curve is momentarily flat (has a slope of 0) at the maximum point, then it curves downward again, its slope reversing.

As rate of change is equivalent to a tangent drawn to the curve, it is easy to spot maximum or minimum points—it is those places at which the curve has a slope of zero and then reverses its slope. This knowledge makes it possible to find the changes of direction—all local maximum and minimum points between boundaries—without drawing the graph. Where the function differentiates to zero, the tangent to the curve is parallel to the axis.

Differentiation is also useful for working with all of the many phenomena that exhibit exponential growth or decay—such as population growth, or radioactive decay. By examining the rate of change at given moments, it's possible to extrapolate to find values for the future (or past).

WAVE FUNCTIONS

The ability of calculus to determine maxima and minima has made it especially valuable for working with all kinds of waveform, from acoustics to optics, from electromagnetism to seismic activity. The earliest work in this field was carried out by the English mathematician Brook Taylor (1685–1731), who in 1714

produced a mathematical description of the vibrational frequency of a violin string. The French mathematician Jean Le Rond d'Alembert (1717–83) refined the model in 1746 to take account of more conditions and limits, and of variation in some properties along the length of the string. His demonstration had two waveforms traveling in different directions. The Scottish physicist James Clerk Maxwell (1831–79) found the same three-dimensional wave when exploring electromagnetism. It enabled him to predict the existence of radio waves. Radio, television, and radar are all developments dependent on the early analytic work on the waveforms of musical instruments.

Further work on the propagation of sound by the Swiss mathematician Leonhard Euler found a trigonometric series at the heart of the problem (1748). In 1822 the French mathematician Joseph Fourier (1768–1830) also found a trigonometric series defining the way heat spreads along a metal rod. From this he developed Fourier analysis, which enabled him to find the values needed to model heat spread for any initial temperature distribution. Fourier analysis is used to analyze complex, composite waveforms, breaking them down into their components and values. For instance, an audio signal can be analyzed into its different

Nicknamed "Dafty" at school in Edinburgh, Scotland, James Maxwell produced work to rival any great physicist.

LEONHARD EULER (1707–83)

The Swiss mathematician and physicist Leonhard Euler spent most of his life in Germany and Russia. He published more

than any other mathematician has ever done, his work filling 60 to 80 volumes. He worked in many fields, making significant breakthroughs not just in analysis, but in graph theory, number theory, calculus, logic, and several branches of physics. He established much of the notation used now, including $f(x)$ for a function of x, the notation for the trigonometric functions, the use of the symbols e (e is sometimes called Euler's number) and i and $\sum$ (for summations). He also popularized (but did not originate) the use of the Greek letter π. One of his most startling discoveries was Euler's identity.

$$e^{i\pi} + 1 = 0.$$

frequencies and amplitudes. Although his methods were not rigorous, they were later refined and are, in essence, used today—for compressing sound into downloadable MP3s, for example.

TOO HARD

Some problems proved intractable even with the use of calculus. The movement of the planets in the solar system, for example, is too complex to be accounted for by straightforward series. The field of dynamic system theory has developed to tackle such problems. Essentially, local data drawn from particular sites within a much larger field are analyzed and results from these are applied to known global properties of

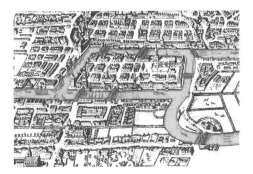

The "Königsberg bridges" problem was solved by Euler in 1735. It asks whether we can cross each of the seven bridges in Königsberg only once, returning to the starting point. By counting the bridges around each area, Euler proved that it is not possible. His proof is now regarded as the first theorem of graph theory.

GREENHOUSE GASES

Fourier was the first person to suggest, in 1827, that gases in the atmosphere may lead to increasing temperature on a planet—the greenhouse effect.

King Oscar II who offered a prize for determining how stable the solar system was.

Newton had used the inverse square law of gravitation to demonstrate the elliptical orbit of planets that Kepler had noticed, but he also found that the system was too complex to calculate if more than two bodies were involved. The king now wanted a solution involving nine bodies—the sun and the eight planets known at the time. Poincaré's solution did not, in fact, deal with nine bodies. He restricted himself to three and even then assumed that one had negligible mass (and so negligible gravitational effect). He modeled a sample of what may happen in a limited area—where the path of a planet intersected with this area—and extrapolated the rate of change to come up with a prediction for the stability of the whole system. Although Poincaré won the prize for his partial solution, he noticed a mistake in his solution and spent more than the prize money in reprinting his solution.

the whole system. Today, computers analyze, approximate, and assess solutions created in this way.

Dynamic systems theory was first developed by Henri Poincaré (1854–1912) for a competition. In 1885, King Oscar II of Sweden and Norway offered a prize to determine the stability of the solar system—saying whether it would continue in much the same state or whether, for example, a planet could fly off on a rogue journey of its own, perhaps colliding with the sun.

Henri Poincaré was blessed with a formidable memory and was able to master many disciplines.

From the end of the 18th century, mathematicians were more willing to accept

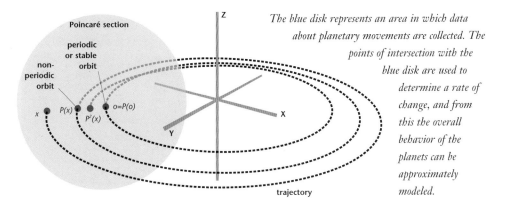

The blue disk represents an area in which data about planetary movements are collected. The points of intersection with the blue disk are used to determine a rate of change, and from this the overall behavior of the planets can be approximately modeled.

SOAP BUBBLES AND ARCHITECTURE

The blind Belgian physicist Joseph Plateau (1801–83) studied the films and bubbles created by soap solution. Soap solution forms minimal surfaces—the minimal surface area that can cover a space. Minimal surface mathematics is a productive area of research. The West German pavilion at the Expo 1967 World's Fair in Montreal, designed by Frei Otto, was based on minimal surface studies of soap films.

---

complex numbers and Gauss began applying the principles of analysis to them in 1811. Analysis using complex numbers—complex analysis—is possible because complex numbers are deemed to follow many of the same laws as real numbers.

Modern analysis differs in many regards from early analysis. Mathematicians discovered that many functions could not be integrated, or behaved in a bizarre way if integrated. As a consequence, integration was redefined by the French mathematician Henri-Léon Lebesgue (1875–1941) around 1900. Instead of taking thin slices of the graph vertically beneath the curve, Lebesgue suggested taking thin slices horizontally. This greatly increased the usefulness of integral calculus as it could now be used with discontinuous functions. It expanded the possible applications of Fourier analysis.

There are very many different branches and applications of analysis and they spill over into all areas of science.

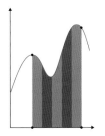

Traditional integration, taking vertical slices

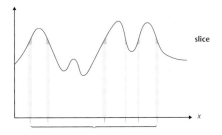

Lebesgue's method, taking horizontal slices

The Lorenz attractor, a model of the movement over time of a chaotic system (here, weather). Even though changes seem random, the overall system forms a pattern.

LOST GLORY

The Japanese researcher Yoshisuke Ueda discovered a chaotic system in the same year as Lorenz (1961), but his supervisor did not believe in chaos and would not let him publish his findings until 1970.

CHAOS THEORY

Poincaré's method is the foundation of chaos theory, which developed greatly during the 20th century. It is a method that enables useful data to be drawn from apparently random systems. Computers have made the study of chaos and chaotic systems possible. The work involves carrying out the same calculations again and again

THE BUTTERFLY EFFECT

A popular encapsulation of chaos theory is the idea that the movement of a butterfly's wings may cause or prevent a tornado, as the small local effect is amplified as it triggers or prevents other changes in the atmosphere. It is likely that the idea comes from a science fiction story by Ray Bradbury, "A Sound of Thunder" (1952), in which a time traveler causes subtle changes in human history by inadvertently killing a butterfly during a visit to Jurassic times.

middle of the sequence. He found the resulting weather prediction was radically different from the one he had obtained the first time. The reason, he found, was that his printout rounded figures to three digits (from six) and this small error was enough to produce a hugely different result.

Chaos theory is applied to many areas of science, including physics, medicine, tectonics, computing, studies of lasers, and electricity. It also has applications outside science in areas as diverse as economics, psychology, and sociology.

with different values—this would be virtually impossible without computers.

A system that appears to be chaotic (in the usual sense of the word) in fact follows strict rules. However, the system is so sensitive to tiny changes in the starting point of its variables that its behavior is, to all intents and purposes, unpredictable. Weather forecasting is notoriously difficult beyond the very short range because a very large number of factors can affect the weather and the outcome is very sensitive to starting conditions. It is effectively impossible to produce an accurate forecast beyond a few days.

It was while working on weather forecasting in 1961 that Edward Lorenz (1917–2008) made a significant discovery in chaos theory. He wanted to repeat a weather modeling operation, but to save time he input figures from a printout he had run earlier, starting his model from the

MOVING ON

The techniques of calculus and analysis are useful in examining trends in data of many kinds. Yet before data can be examined, they must be collected and processed. Surprising as it seems, the idea of collecting data on which to base decisions—and making that data collection both rigorous and fair—is a relatively recent development. The branch of mathematics that has come to be known as statistics has grown up over only the last 400 years. Interestingly, its emergence coincided with the development of calculus, which has become an important tool in statistics and probability studies.

CHAPTER 7

NUMBERS
at Work and Play

Calculus and analysis are a long way from the everyday encounters with numbers that many of us have. Even though most of the science we come into contact with, most of the products we use, and much of the world around us depends on activities in higher mathematics, our everyday encounters are more likely to lie with statistics and probability. In finance, gambling, games, the economy, and many other spheres, numbers as predictors and risk assessors help us to make decisions—whether about buying a lottery ticket, taking out life insurance, or flying on a plane.

Numbers, and the possibilities they offer, are with us all the time.

Cheer Up, It May Never Happen

Humankind has played games of chance for millennia. This is playing with numbers; the fall of the dice or roll of the roulette wheel are effectively random, and winning at these games demands either large slices of luck or great proficiency in calculating probabilities and risks.

Very simple probabilities are easy to see—if we toss a coin, there is a 1 in 2 chance that it will land heads and the same chance that it will land tails. If we toss a coin a large number of times, we will probably get about as many heads as tails. This was first noted by the Swiss mathematician Jacob Bernoulli in a treatise published posthumously in 1713. He did acknowledge that the result is so patently obvious that even a very stupid person would notice it, but he is still given credit for it as he spent 20 years developing a rigorous demonstration of why it is true. He called it his Golden Theorem, but it is generally known now as the Law of Large Numbers. Casinos depend on it; although an individual gambler may have a run of good luck, over time a casino can expect to keep 5.3 percent of all the money bet on a roulette wheel.

Although it is possible to "beat the bank" over a short period of time, the casino is fairly certain to win in the long run.

Between the obvious probabilities and the Law of Large Numbers, problems of probability become more complex. What are the chances of getting tails exactly five times in a row? If we throw three dice, what is the chance of getting three sixes?

We need to do a little work with probability to be able to calculate these; the chance of getting tails five times in a row is 1 in 2^5 = 1 in 32; the chance of throwing three sixes is 1 in 6^3 = 1 in 216.

For most of the many thousands of years that people have been playing games of chance, they had no way of working out the probabilities of different outcomes beyond the few that are very obvious or for which it is easy to enumerate the possibilities.

DICE AND CHAOS

Although the fall of dice or spin of a roulette wheel are effectively random, they are actually determined events. The starting position and all prevailing conditions—including the direction and force of the throw, the surface of the table, and the exact features of the dice—will determine the outcome. However, there are too many conditions, and their measurement is too difficult, for the outcome to be modeled or calculated.

A GAME OF CHANCE

Probability—the chance or likelihood of an event happening—entered mathematics in the 17th century and it was in the context of a game of chance. Although Gerolamo Cardano had written on games of chance in the 1520s (see page 132–3), his work was not published until 1633 so he lost out to Fermat and Pascal. In a series of letters, the pair discussed a problem proposed by a gambler, the Chevalier de Méré:

Two players are playing a game of pure chance on which each has bet 32 coins. The first to win three times in a row claims the pot. However, their game is interrupted after only three games. Player A has won twice and player B has won once. How can they divide the pot fairly?

The two mathematicians both came up with a 3:1 distribution in favor of player A, though they arrived at the solution by different methods.

Fermat gave his answer in terms of probabilities. Two more games is the most that would be needed to decide the match, and there are four possible outcomes AA, AB, BA, BB. Only the last would make B the overall winner, so he has a one in four chance and should receive a quarter of the winnings. Pascal proposed a solution based on expectation. Assuming B wins the next round, each player would have an equal claim to 32 coins. Player A should receive 32 coins anyway as he definitely has two wins. The chance of B winning this next game is 50 percent so he should have half of the remaining 32 coins. Player A also has a 50 percent chance of winning and should

have the last 16 coins. Again, player A receives 48 and player B receives 16 coins. Pascal's strategy was the one that won approval among mathematicians dealing with chance events.

ALL'S FAIR...

Although games of chance continued to interest mathematicians, another impetus was the legal idea of a fair contract. In a fair contract, the parties have equal expectations. This was an important concept because fair expectations were at the heart of the justification for money-lending. Christian doctrine bans usury—profiting from lending money. To get around the difficulty, lenders were considered to be investors who put in money at their own risk and could fairly expect to share in the profits.

Until the 17th century, the rates for loans and annuities were fixed with no regard for any mathematical concept of risk or how it might be calculated. The first treatise on calculating risk appeared in the Netherlands in 1671, produced by Jan de Wit after consulting Christiaan Huygens. At the time, annuities were sold by the state to raise money, often to finance wars. The return

Jan de Wit realized that risk should govern rates of return.

was always a seventh of the value of the annuity, paid each year until the holder's death. The age or health of the holder was not taken into account. Clearly, without any assessment of how long the state may have to pay the annuity holder, this could be expensive. Even though de Wit could see the flaws in the system, there were at the time no data on mortality at different ages, so little could be done to improve the system—and little was done. It was not until 1762 that the Equitable, an insurance company in London, England, began to price its policies on the basis of calculated risk, or probability.

GOD EXISTS—PROBABLY

Probability did not become an exact mathematical concept until the 18th century, and was still generally considered an indistinct idea based on common sense into the 19th century. The French mathematician Pierre-Simon de Laplace (see page 174) referred to probability as "good sense reduced to calculation."

Interestingly, a link between chance and religion became a central interest of natural theology in the 18th century. John Arbuthnot (1667–1735) produced evidence that God definitely exists from a study of christening statistics in London between 1629 and 1710. He showed that there were slightly more boys born than girls—14 boys christened for every 13 girls—yet by the age of marriage the balance of the sexes was equal. If we assume that the chance of a child being born a boy is 0.5, the chance of more boys than girls being born every year for 82 years is 0.5^{82}. The same pattern of more male births is found throughout

PASCAL'S WAGER

In 1657–8 Blaise Pascal wrote a philosophical essay in which he described the "wager" a skeptic should make. The penalty for not believing in God (the Christian God, for Pascal) could be eternal damnation; however, the cost of believing in God if He turns out not to exist is slight. At most, the person who chooses to believe may relinquish a few fleeting pleasures and spend a few fruitless hours in church. Although the skeptic may feel that the chance of God's existing is very small, the cost of losing the wager is so high and the price of belief so comparatively low that it is a better bet to believe than not believe.

the world. Arbuthnot took this as incontrovertible evidence of Divine Providence at work, setting up society with the perfect balance. (It doesn't seem to have occurred to him that Divine Providence could equally well have killed fewer boys on the path to adulthood, thus avoiding the suffering of bereaved parents at the same time as achieving the required balance.) The argument was generally adopted and refined. However, Nicolas Bernoulli, the more rational Swiss mathematician, suggested that perhaps the probability of a male birth was not 0.5 at all but 0.5169, which would produce exactly the required result with no need for divine intervention.

MAKING DECISIONS

As with Pascal's wager, many decisions that may be influenced by a knowledge of probability are also affected by a more subjective perception of desirable outcomes and the concept known as "marginal utility." Imagine a national lottery, in which tickets cost one ducat (a coin in use in much of Europe in the 18th century) and the prize is a million ducats. For a poor man, a ducat is very valuable, and the payout immensely so. For a rich man, a ducat is of little consequence, though the payout is still valuable to him. The rich man can better afford to bet a ducat than the poor man, but as he has less need of the prize he might not bother. Although the probability of winning is equal for both, the decision about whether to buy a ticket is very different for each.

In the 1750s and 1760s, inoculation against smallpox was a topical subject of debate. The inoculation used live smallpox virus and in a small number of cases produced smallpox (Jenner's vaccine produced from cow pox was a later and safer introduction). Smallpox was very common, often deadly and, even when not fatal, frequently led to lifelong damage such as blindness or brain damage. Someone who did not have the vaccine stood a high chance of contracting smallpox at some time in the future, and a 1 in 7 chance of dying from it. Someone who chose to have the vaccine

stood a small chance (not measured) of dying immediately of smallpox brought on by the inoculation, but otherwise virtually no chance of dying of smallpox in the future. The purely mathematical calculation, carried out by Daniel Bernoulli, suggested that there was only one sensible choice— inoculation. But the French mathematician Jean Le Rond d'Alembert, among others, argued that many people may prefer the better chance of surviving the next week or two to the assurance of safety in the future. (Today, plenty of people prefer the immediate advantage of long-haul flights to the long-term benefit of still having a planet to live on.)

INDEPENDENCE

People are not only affected by marginal utility and the preference d'Alembert noted for short-term benefit. They may also be swayed by superstition that has no grounds in statistical probability at all.

Imagine flipping a coin ten times; the probability of getting heads each time is 1 in 2^{10}. Suppose the first time it is heads. Now the probability of all ten flips being heads is 1 in 2^9. If the first nine come up heads, the probability of ten heads, by the last time, is 1 in 2. Now suppose you want to fly on a plane. You know that the chances of dying in a plane crash are, say, 1 in a million on any particular flight (this is not the real probability). You have already made 1,000 flights

Human beings seem happy to gamble with their long-term future if it means gains in the short term.

HERD IMMUNITY

Some diseases have been completely or nearly eradicated by national inoculation programs. An example is measles, once endemic in the western world but now rare in countries with inoculation programs. However, worries about the safety of the vaccine in the 1990s led to a reduced take-up of childhood vaccination in the UK and measles began to take hold again. While the vast majority of a population has immunity, a few unprotected individuals benefit from the "herd immunity" as the disease can't get a foothold among the inoculated population. However, as the number of unprotected individuals rises, the presence of the disease increases to the point where it can spread among the uninoculated population.

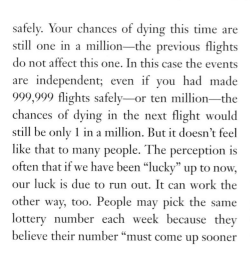

The dilemma facing parents who were unconvinced about the safety of the vaccine mirrored that of the people making a choice about the early smallpox vaccine. For society as a whole, there was a moral dimension—was it right that a few individuals should avoid the (possible) risk posed by the vaccine and depend on benefiting from the herd immunity acquired at the cost of everyone else taking that risk? For mathematicians and medics, there was a different question: what proportion of the population could remain unvaccinated before their safety was compromised?

safely. Your chances of dying this time are still one in a million—the previous flights do not affect this one. In this case the events are independent; even if you had made 999,999 flights safely—or ten million—the chances of dying in the next flight would still be only 1 in a million. But it doesn't feel like that to many people. The perception is often that if we have been "lucky" up to now, our luck is due to run out. It can work the other way, too. People may pick the same lottery number each week because they believe their number "must come up sooner or later." Few people pick numbers 1, 2, 3, 4, 5, and 6 because they believe (irrationally) that this combination is less likely to be drawn than any other. This tendency is not so far removed from the Ancients who believed the number 3 had special properties, or who wore a magic square for protection.

INTERDEPENDENCE

When choosing whether to board a plane, people are dealing with random events—they have no control over whether the plane will crash. A situation that is harder for

John von Neumann was a member of the Institute for Advanced Study at Princeton, a group of academics affectionately known as the "demi-gods."

mathematicians to model is that in which one person's actions are dependent on or linked with those of another person (such as the decision about whether to vaccinate a child). This is addressed by game theory, developed in the 1940s by the Hungarian-American mathematician John von Neumann and the German-American Oskar Morgenstern.

Despite its name, game theory is concerned with the serious pursuits of economics rather than the frivolity of games. Morgenstern and von Neumann saw that the mathematical models developed for systems in physics and other areas of science were poor tools for working with economics and other studies that involve human behavior because they were based on the actions of disinterested parties. When people make choices, they try to maximize

the benefit for themselves. They may also try to minimize the detriment to others—or they may pay no attention to the impact on others, or even act to spite them.

Game theory tries to take account of the motives and insights of people acting in the situation that is modeled, as well as many other relevant aspects. For example, players—which may be individuals, groups, nations, or corporations, for example—may be in direct competition or may cooperate to a greater or lesser degree. They may be competing for a finite resource or infinite resources. They may be in full possession of all relevant information, including the actions of other players, or have only partial access to information. There are different game theory models to cover these and other possibilities. Game theory often produces a matrix of outcomes, which can then be analyzed.

BACKWARD REASONING

Proofs such as that of Arbuthnot that God exists work backward from effects to causes—there are equal numbers of marriageable men and women, therefore God exists. Jacob Bernoulli demonstrated that, if the probability of an event is not known, it can be inferred from looking at the results of experiment or observation as long as the observer has sufficient knowledge and experience. He gave as an example the fact that if a coin is tossed enough times, the ratio of heads to tails approaches ever more closely the ideal 1:1. A formal demonstration of probability in this way was made independently by Thomas Bayes and by Laplace and is now known as Bayes' theorem. Laplace famously used it to argue the probability of the sun rising tomorrow, given our knowledge that it has risen every day for the last 6,000 years (which in 1744 was considered to be the age of the Earth).

Laplace and his contemporaries tried to put probability at the heart of the moral sciences, though their attempt was somewhat dubious. Enlightenment philosophers and reformers were concerned with the value of the judgments made by electorates and juries—would they reach the right decision or elect the best candidate? They addressed this as a problem in probability. Assuming that each juror acted independently (French juries did not deliberate) and had a greater than 0.5 chance of reaching the right verdict, they worked out the optimum size of jury and the majority needed to reach a safe conviction. The practice of deciding jury size and majority using probability continued until the 1830s. By then the system was coming into disrepute and a pupil of Laplace, Siméon-Denis Poisson, used new statistics to produce a better model.

Before probability could be used effectively in any area, though, reliable information was necessary. Statistics and probability go hand in hand.

PIERRE-SIMON, MARQUIS DE LAPLACE (1749–1827)

The French scientist and mathematician Pierre-Simon de Laplace was most famous for his work on astronomy and his application of probability to scientific problems. He was the son of a peasant farmer, who revealed mathematical ability while at a military academy in Beaumont. In 1766 he went for one year to the University of Caen, but left for Paris, where Jean d'Alembert helped him to secure a professorship at the École Militaire. He taught there until 1776.

Laplace applied Newton's theory of gravitation to the movement of the planets. He perfected the contemporary model of the solar system and demonstrated that apparent changes are not cumulative, but occur and correct themselves in predictable cycles. (Isaac Newton had suggested that divine intervention was sometimes needed to put the solar system right!) Laplace was the first to suggest that the solar system was formed by the cooling of a vast cloud of gases.

His explanation of planetary motions made him a celebrity. Laplace was president of the Board of Longitude, helped to organize the development and introduction of the metric system, and for six weeks was minister of the interior under Napoleon.

Samples and Statistics

Without information on which to base decisions, it is possible to calculate only the most basic probabilities. Astonishingly, it was not until the late 17th century that people began to recognize the true value of collecting numeric information about populations and economies. Suddenly, statistics were everywhere and computing with them gave new insights into how societies might work. For the first time, the guesswork was taken out of planning and the burgeoning science of statistical analysis had material to work with and aims to work toward.

PEOPLE COUNTING

Collecting information about the number of people living in an area by taking a census has been practiced intermittently for thousands of years. The Babylonians, Ancient Chinese, Egyptians, Greeks, and Romans all held population censuses. In Christian tradition, the parents of Jesus traveled to Bethlehem immediately before His birth because the five-yearly census required everyone in the Roman Empire to return to their place of birth to be counted.

The very basic information collected in these early censuses allowed rulers to work out how much money could be collected in taxes, how many people could be recruited for an army or building project, and how much food could be produced or would be needed. In Egypt, it was also used to redistribute land after the annual flooding of the Nile. But no additional analysis of population data was carried out and only the most basic details were collected. Often, the census data were not reliable. If people

expected to be taxed on the basis of how many lived in a house, a few might be missed out, for example.

In 1066, after the conquest of Britain by Norman invaders, William the Conqueror held a thorough audit of his new lands. This included a census and a listing of every item of property in the land. It was written up in the *Domesday Book*—a massive undertaking for the 11th century and one that still provides valuable statistics for historians. Thereafter, there was no enthusiasm for regular census-taking. Although bishops in many parts of Europe were supposed to keep count of the families in their dioceses, there was little information about population levels. Some people even believed that taking a census was sacrilegious, citing a story from the Bible in which King David attempted a census which was interrupted by a terrible plague and never completed.

The first regular census in modern times was carried out in Quebec, Canada in 1666. In Europe, Iceland was the first in 1703, followed by Sweden in 1749. The United States held its first ten-yearly census in 1790 and the UK in 1801; the United States had just under 4 million inhabitants and the UK 10 million (previous estimates had put the UK population at between 8 and 11 million).

THE RISE OF STATISTICS

In 1662, the English statistician John Graunt published a set of statistics drawn from mortality records in London, and in the 1680s the political economist William Petty published a series of essays on "political arithmetic," which provided statistical records with calculations—some

THE CENSUS AND COMPUTERS

The demands of census-taking were a considerable spur to the development of technological aids to calculating. The first machine for working with census data was used in 1870. Census data were transcribed on to a rolling paper tape displayed through a small window. In 1884 Herman Hollerith (1860–1929) acquired the first patent for storing data on punched cards and organized the health records for Baltimore, Maryland, New York City, and New Jersey, which won him the contract to tabulate the 1890 census. The huge success of this census opened other markets to Hollerith and his machines were used in Europe and Russia. He incorporated his Tabulating Machine Company in 1896, which later became IBM.

Hollerith produced a mechanical tabulator based on the idea that all personal data could be coded numerically.

quite bizarre, such as the monetary value of all people in Ireland. On the whole, governments encouraged or financed statistical surveys and guarded the results jealously, using them to increase the power of the state. They were still inextricably tied up with superstition and followed very unscientific methods. One of the most famous "political arithmeticians" was the Prussian Johann Süssmilch, who published three volumes over more than 20 years, ending in 1765, proving again the existence of God revealed in the harmony of social statistics.

Other statistics were collected by scientists, professionals of different types, and humanitarians. Indeed, there was a growing enthusiasm for statistics, which became something of a mania during the early 19th century. Suddenly, everything was studied, counted, audited—the weather, agriculture, population movements, the tides, the land, the Earth's magnetism… The European countries that had empires surveyed their new acquisitions and took censuses in their colonies. As Americans moved westward, claiming more land, they charted it and logged its resources.

SOCIETY IS TO BLAME

The Belgian mathematician Adolphe Quetelet (1796–1874) was a champion of statistics as the basis of the social study that he termed "social physics." He examined data of all kinds, using the techniques common in some scientific disciplines of amassing a vast collection of data and looking for emergent patterns. To his surprise, he found them everywhere, not just in the areas where Divine Providence

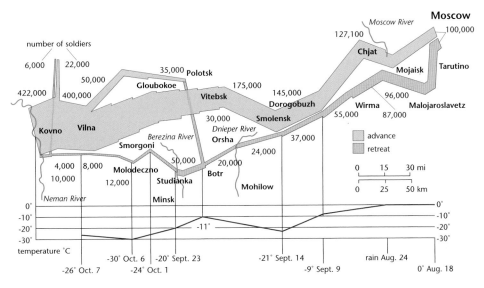

might be expected to operate. In particular, he was impressed to find that crime figures followed a predictable pattern. He conjectured that they are a product of society rather than individuals and that, while an individual criminal may be able to resist the urge to commit a crime, the overall pattern of crime rates is altered little by individual actions. He felt that the proper study was of crime rates rather than criminals and that the proper remedy to crime lay in social action, including education and an improved judicial system. Careful use of statistics to examine the effects of changes and suggest directions for future change would, he felt sure, produce the desired results.

Quetelet's thesis prompted some debate on the apparent conflict between statistics and the doctrine of free will—if crime rates can be determined by statistical methods and are unchanging over time, how much freedom do individuals really have over their actions?

One of the most accomplished graphical representations of statistics ever made is Charles Minard's graph of Napoleon's disastrous campaign in Russia in 1812. It shows mortality on the way to and from Moscow and correlated with temperature. The width of the green and orange lines represents the size of the army, showing how it dwindles. Only 4 percent returned from the campaign.

STATISTICS MEET SCIENCE

Perhaps surprisingly, it was not until the middle of the 19th century that statistics began to be applied to science with the same enthusiasm and rigor as they had been applied to social science. In the 1870s the Scottish physicist James Clerk Maxwell often explained his theory of gases with reference to social statistics. From the very large number of random movements of molecules he derived thermodynamic laws—order from chaos. He argued that, just as statistics relating to crime or suicide can yield consistent results from the unordered acts of individuals, so predictable

FLORENCE NIGHTINGALE (1820–1910)

Florence Nightingale enjoyed a privileged childhood in England, where her father taught her languages, philosophy, history, and mathematics. She claimed to have had a message from God telling her she had a vocation and later wanted to train as a nurse. Her family resisted and she became instead an expert on public health. She did later train as a nurse and, during the Crimean War (1854–56), was put in charge of the hospital at Scutari, in Turkey, where she revolutionized healthcare for wounded soldiers. She kept copious notes and after the war put together an extensive report from the statistics she had gathered. She used innovative ways of presenting information in graphs, such as the "coxcomb" graph (above right).

Nightingale worked tirelessly to improve conditions in the British army. She founded the first training school for nurses anywhere in the world, the Nightingale School for Nurses in London, and established the professional footing of nursing.

Causes of mortality in the army in the East, April 1854 to March 1855

- non-battle
- battle

Nightingale was a pioneer in the analysis and presentation of statistics. "Coxcomb" graphs were designed to be understood by everybody.

outcomes on a large scale could be extracted from acts that are unpredictable on the small scale. But before statistics could be applied, it had to develop as a mathematical discipline. Mathematical methods specifically applying to statistics began to emerge from the end of the 18th century and proliferated rapidly.

> "[Statisticians] have already overrun every branch of science with a rapidity of conquest rivalled only by Attila, Mohammed and the Colorado beetle."
>
> Maurice Kendall, 1942

Statistical Mathematics

WHAT'S NORMAL?

Abraham de Moivre (1667–1754) was the first person to notice the characteristic bell curve of the normal distribution (see below). The curve plots the frequency or probability of values against the values themselves. The most frequently occurring results are at the top, representing the mean value; the results that deviate most from this norm and occur least frequently are on the lower arms of the curve. The slope of the curve is determined by the degree of variation within the sample. Approximately 68 percent of the values in the normal

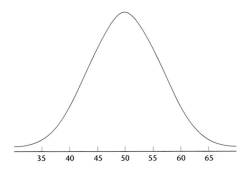

distribution are said to fall within one standard deviation of the norm.

The normal distribution curve and the concept of standard deviation from the norm were widely used to assess statistics in many different fields. Laplace used the model, too, in his probability studies, particularly in applying probability to very large numbers of events. Quetelet argued that virtually all human traits conformed to the normal distribution curve,

Abraham de Moivre was a pioneer in analytic geometry and the theory of probability, being first to notice the normal distribution curve.

from physical attributes such as height to characteristics of psychological profiling such as propensity to get married or commit suicide.

WORKING WITH ERROR

The early 19th century saw a rapid rise in mathematical methods involving statistics. Work on measuring the Earth's longitudinal circumference in order to determine the length of a meter (to be 1/40,000,000 of the circumference) needed statistical methods to deal with errors and inconsistencies in geodetic measurements. In 1805, the French mathematician Adrien-Marie Legendre (1752–1833) proposed a technique that has come to be known as the "least squares" method. He took values that minimized the sums of the squares of deviations in a set of observations from a point, line, or curve drawn through them. Gauss became interested in the method and showed in 1809 that it

Adrien-Marie Legendre has a crater on the moon named after him.

179

METHOD OF LEAST SQUARES

The method of least squares calculates the best line through a set of points by working out the smallest possible sum of the squares of deviations from the line of all the points. Squares are used to remove the difficulty of dealing with both positive and negative deviations, since when squared they will both give a positive result.

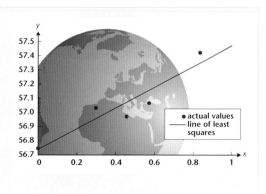

gave the best possible estimate if we assume that the errors in measurement follow the normal distribution. The method of least squares was applied to statistics in all fields and became the principal tool of statisticians in the 19th century. It was often used to estimate whole populations from a study of a small sample.

PERFECTING HUMANITY

Francis Galton, a cousin of Charles Darwin, took an interest in the variation highlighted by normal distribution and standard deviations.

He used a model, known as the Galton board, to show how a normal distribution is achieved (see below). A set of pegs is arranged in a triangle above a row of cups. Ball bearings dropped at the top of the triangle bounce down through the pegs to fall into a cup. A few fall into outlying cups but most fall into the cups in the middle of the board, forming a normal distribution curve.

Galton applied statistical ideas to heredity to show how variation tends to be bred out, and generations of an organism tend to revert to similar levels of variance.

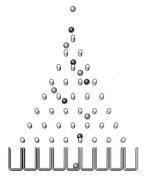
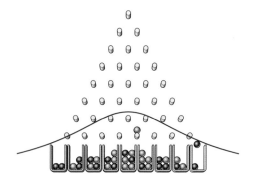

Ball bearings dropped on to the Galton board at the top are deflected into the cups at the bottom. The distribution of ball bearings in cups demonstrates a normal distribution curve.

So although the children of exceptional parents may be exceptional themselves, at least in some ways, on the whole they tend to regress toward the general population as a whole. Galton took his research in an alarming direction, becoming the founder of the eugenics movement, which aimed to guide human evolution toward perfection. He wanted to breed in "good genes" in the way that breeders select the best genes in farm animals and crops.

Although originally he was interested primarily in genetics and heredity, Galton recognized the application of his statistical methods to other areas and stressed the adaptability of the tools he developed.

COURTING RANDOMNESS

Developments in statistics aimed to enable information from a small sample of data to be extrapolated or applied to a larger population. By deciding the rate of crime or marriage or an inherited disease in a sample of the human population, for example, researchers hoped to reach conclusions about the rate in the whole population. The results of any statistical survey depend, of course, on the quality of the sample measured. The head of the Norwegian Central Bureau of Statistics, A. N. Kiaer, aimed to draw samples that covered the full range of representative variables in the population, such as old and young, rich and poor. The English statistician Arthur Bowley was one of the first to try to introduce randomness into sampling. The Polish statistician Jerzy Neyman brought these two concerns together in 1934, trying to ensure that a sample included representatives of major variables but that

the individuals included should be chosen at random. The first triumph of this technique of stratified sampling came in 1936 when George Gallup's poll predicted the re-election of Franklin D. Roosevelt in the United States, while a larger, unstratified sample, confidently (and wrongly) predicted the opposite result. Gallup drew on a sample of only 3,000 voters, while Literary Digest, the opposing pollsters, polled 10 million. Roosevelt won with the largest landslide in American history. A large sample is no guarantee of a representative sample or an accurate result.

Experimental design went hand in hand with the development of statistical tools. The use of a control group to compare with

The landslide reelection of Franklin D. Roosevelt in 1936 came as no surprise to Gallup, who had used stratified sampling to predict such a result.

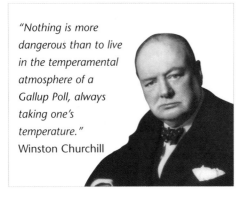

> *"Nothing is more dangerous than to live in the temperamental atmosphere of a Gallup Poll, always taking one's temperature."*
> Winston Churchill

theory of evolution, which had been thrown up by the experimental work on inheritance of the Austrian botanist Gregor Mendel. He developed the method—which now seems ridiculously obvious—of varying only one condition in an experiment at a time and comparing results with a control group. Although earlier experimenters had done this to some degree, it was felt to be immoral where human subjects were concerned, and so rigorous use of control groups and random allocation of individuals to the control or experimental group had not been practiced previously. Fisher also advocated repeating experiments and looking at the

the experimental group, and the random allocation of individuals to the control or experimental group, emerged as standard procedure during the early years of the 20th century. In particular, the British geneticist and statistician Sir Ronald Aylmer Fisher (1890–1962) reshaped experiment design in many fields, including psychology, medicine, and ecology, in the years after World War II. He began his research in genetics, where he used statistical analysis to reconcile inconsistencies in Darwin's

The lavarand random number generator developed by Bob Mende in 1996 produced random numbers using a computer program seeded with digital photographs of the patterns produced by lava lamps.

THE DIFFICULTY OF BEING RANDOM

It is not only in sampling that randomness is actively sought. In games of chance for high stakes, there is an imperative to make sure that events that are supposed to be random are in fact just that. Cryptography also demands randomly generated numbers. This is much harder than it at first appears. As chaos theory demonstrates, many events that look random are actually not, but are governed by complex laws and a large number of variables.

The systems used to pick numbers for large-scale gambling ventures such as national lotteries are very carefully designed and engineered to remove, as far as possible, all bias in the selection methods. It is very difficult to produce computer algorithms for picking random numbers, so most lotteries use mechanical methods instead. (These also have the advantage of looking more spectacular than computers.) Computers that can generate genuinely random numbers do so using a physical source such as atmospheric noise (e.g., www.random.org).

variation in results to determine the margin of error. The most influential statistician of the 20th century, Fisher summed up his findings in the highly influential text *Statistical Methods and Scientific Inference* (1956). One of Fisher's most important developments was in the analysis of variance (called ANOVA) which looks at the points in a sample that vary from the norm. It is used to assess whether or not results are statistically significant— that is, whether they are likely to reflect a real trend, change, or cause, or whether they could have come about by chance.

British mathematician John Conway's "Game of Life" quickly gained a cult following by simulating life, death, and change in a "society of living organisms."

COMPUTERIZATION

The burden of calculating with very large sets of data has been made easier by the widespread use of computers. While earlier statisticians had the laborious task of carrying out calculations for each data point by hand, their modern counterparts can feed all their data directly into a computer and leave it to apply the necessary statistical tools and provide the analysis and graphs. Often, the data are even collected by computers directly from sensors. We can now handle immense data sets, so large that they could not have been handled in a whole lifetime without computers. It means that statistical analysis can be applied in all areas of life, determining patterns and projecting outcomes in areas as diverse as the effect of

early education on crime rates, the likely spread of epidemic disease, and the effects of global warming.

A famous illustration of the importance of initial conditions is John Conway's "Game of Life" (1970). This is a cellular automaton—a computer simulation of an evolving population or universe in which an initial organism or automaton makes copies of itself, which succeed or fail according to various conditions (such as overcrowding, lack of resources, etc.). Conway created it in response to a problem presented by John von Neumann in the 1940s relating to constructing a machine that could make copies of itself. The "Game of Life" is not a game in the usual sense of the word, in that there are no active players. After the

SETI@HOME

The SETI project—Search for Extra-Terrestrial Intelligence—collects radio data from space on a continual basis, and is starting to look also for pulses of laser light. Its stated aim is "to explore, understand, and explain the origin, nature, and prevalence of life in the universe." SETI's task is to examine the constantly growing data set for patterns that might indicate a deliberate radio transmission. To do this, it asks volunteers around the world to install a screensaver that imports chunks of data from SETI over the Internet and processes them on the computer while it is not being used. In this way, SETI makes use of millions of hours of free computer time on personal computers around the world. Each PC reports its results back to SETI and any possible patterns are flagged for further investigation. An unimaginably large task in statistical analysis is being carried out at very little cost and much more quickly than it could be managed using dedicated computers.

The SETI equation

The Drake equation (1961) is suggested as a way of calculating the likely number of planets that have intelligent life in the Milky Way:

$$N = R^* \times fp \times ne \times fl \times fi \times fc \times L$$

where

N = the number of civilizations in the Milky Way whose electromagnetic emissions are detectable.

Looking for signs of life: radio antennae that form part of the Very Large Array astronomical observatory in New Mexico, USA.

R^* = the rate of formation of stars suitable for the development of intelligent life.

fp = the fraction of those stars with planetary systems.

ne = the number of planets in each solar system with an environment suitable for life.

fl = the fraction of suitable planets on which life actually appears.

fi = the fraction of life-bearing planets on which intelligent life emerges.

fc = the fraction of civilizations that develop a technology that releases detectable signs of their existence into space.

L = the length of time these civilizations release detectable signals.

"Nothing in the universe is unique and alone, and therefore in other regions there must be other Earths inhabited by different tribes of men and breeds of beasts." Lucretius, 50 BCE

instigator sets up initial conditions, the game runs, producing generations that flourish or perish according to the consequences of the starting conditions. The original game used populations of colored squares in a grid, but it spawned a whole industry of computer simulation games, some of them immensely complicated, that produce populations of creatures or other entities. The interest in cellular automata that grew out of Conway's game has found applications in many fields, including research into human, animal, and viral populations, growth of crystals,

economic problems, and many other areas in which complex patterns develop organically.

Moving On

Much of the work on statistics in the last hundred years or so has led to analysis of groups or sets of data in quite complex ways. The behavior of sets—whether of numbers or anything else—is the subject of set theory, first developed in the second half of the 19th century. The appearance of set theory has been one of the most important developments in the history of mathematics.

CHAPTER 8

The Death of
NUMBERS

Analyzing data gathered from populations, experiments, and other sources leads to a search for patterns that can be used to categorize and group items. The natural result of this is to divide items into sets and compare these. Everything in the universe can be defined by its membership of sets.

Relationships between these sets gives further information about the objects. When mathematicians turned to set theory at the end of the 19th century, they found a rich lode that would provide methods for dealing with everything.

As set theory developed, it built a logical language of its own that was eventually turned back on itself. Set theory could be used, it appeared, to explore mathematics, to prove mathematical theorems, and even to dissect and analyze set theory. It allowed mathematics to encompass everything and yet to become so abstract and abstruse that it apparently dealt with nothing.

SETI (page 184) aims to identify a subset of planets that host life.

Set Theory

At the very beginning of the history of numbers, before humankind began counting, it is likely that people compared sets of objects—are there enough spears for hunters to have one each? Are there more or fewer sheep in the pen than pebbles in the tally pile? As the requirements of humankind became sophisticated, mathematics moved away from these sets of objects, developing a concept of number that could be applied universally. Now, thousands of years later, mathematicians have returned to sets, but with a new insight—the possibility of an infinite set.

ORIGINS OF SET THEORY

Set theory was developed by Georg Cantor between 1874 and 1879. He defined a set as a collection of definite, distinguishable "objects of perception or thought" that were conceived as a whole entity. So there is a set of positive integers which can be thought of as an object in its own right. But there is also a set of people who are employed as firemen, a set of molecular structures that form hydrocarbons, and so on. Although

---

INSULTS FLY

Among the insults hurled at Cantor and set theory were that it was a "grave disease" infecting mathematics (Poincaré); "utter nonsense," "laughable," and "wrong," and that mathematics was "ridden through and through with the pernicious idioms of set theory" (Wittgenstein); and that Cantor was a "scientific charlatan," "renegade," and "corrupter of youth" (Kronecker).

---

the basic principle is extremely simple, logical thought about sets soon leads into complicated concepts that blur the boundary between mathematics and philosophy. Early critics argued that set theory dealt only with fictions, not with anything that reflected reality, that it violated the principles of religion and that it was not mathematical. It is true that set theory is a branch of pure mathematics that may seem to have little application to the everyday world. It has proved immensely valuable, however, in enabling manipulation of complicated mathematical concepts. Set theory is itself capable of being defined, analyzed, and refined by applying the logic of sets to its concepts.

SETS FOR BEGINNERS

The fundamental concept of sets is very simple. Any group of objects or numbers, whether or not they have real or enduring existence, forms a set. Any individual member of a set may be a member of many sets. Sets overlap, and some contain other sets (subsets). A set may have an infinite number of members, in which case it is an infinite set.

Arithmetic with sets is not quite equivalent to arithmetic with numbers. If two sets are added together (called the union of sets), the new set comprises all the members of both sets but with no duplicate entries. The intersection of two sets comprises the members that are in both sets. A set with no members is a null set, designated by Ø.

In general, the order of elements in a set is not relevant. So while the coordinate pairs (x, y) and (y, x) are not the same,

GEORG FERDINAND LUDWIG PHILIPP CANTOR (1845–1919)

Georg Cantor was a German mathematician who is best known as the creator of set theory. His parents were Danish, but settled in Frankfurt in 1856. Cantor's mathematical ability became obvious during his early teens. He went to university in Berlin and Zurich to study mathematics, philosophy, and physics. He was taught by Weierstrass, whose influential work on analysis impressed Cantor. His PhD thesis, submitted after a single term, had the title *In mathematics the art of asking questions is more valuable than solving problems.* He became a professor at Halle in 1879.

Cantor worked first on the theory of numbers, then turned to the theory of trigonometric series and extended the work of Bernhard Riemann. He met Richard Dedekind (1831–1916) while on his honeymoon and the pair became lifelong friends. Their exchange of letters brought Cantor to his most important work, on sets and the concept of transfinite numbers. He faced considerable opposition from Leopold Kronecker (1823–91), who blocked publication of his work and his advancement within the university. Kronecker had previously been Cantor's mentor, but now believed Cantor's work to be meaningless and the transfinite numbers he was writing about to have no existence.

Opposition came from the Church, too, which felt his work challenged the unique infinity of God. This opposition made Cantor particularly sympathetic to young scholars later in life, whom he helped and encouraged in the face of an entrenched and resistant system.

Cantor's work established set theory as a branch of mathematics and laid the groundwork for much of the development in mathematics in the 20th century.

the sets {x, y} and {y, x} are identical.

Cantor's definition of a set was that it is a grouping into one entity of objects of any type, though each object also retains its own identity. For any object x its relation to a set A is always that it is or is not a member—shown by $x \in A$ or $x \notin A$.

The condition $x \in A$ is true only if x meets the conditions of the formula

> *"Nowadays it is known to be possible, logically speaking, to derive practically the whole of known mathematics from a single source, The Theory of Sets."*
>
> Nicolas Bourbaki, 1939

BERTRAND RUSSELL (1872–1970)

Bertrand Russell was a British mathematician and philosopher, born into an aristocratic family. He was orphaned at six and brought up by his grandmother, home-tutored and isolated from other children. He went to Trinity College, Cambridge to read mathematics, but changed after a short time to philosophy. Much of his philosophical work was on the philosophy of mathematics.

He was influenced by Weierstrass, Dedekind, and Cantor, who all wanted a formal, logical basis for mathematics. Russell aimed to prove with *Principia Mathematica* that mathematics was nothing but logic. However, he discovered a paradox that he simply could not resolve except by redefining the basis of logic: a barber says he shaves everyone who does not shave

himself. Who shaves the barber? Logicians have found different ways of adapting set theory to deal with the paradox, which attempt to make the definition of a set more restricted and precise.

S(x) which defines members of the set. This is the principle of abstraction. That a set is defined by its members is the principle of extension. The number of elements in a set is called its cardinal number—the set {4, 5, 6} has a cardinal number of 3. The cardinal number of a set A is written $\bar{A}$. Any subset of a finite set has a smaller cardinal number than the original set. If we imagine a set of "all cars," there are clearly fewer members in the subset "red cars."

WORKING WITH INFINITY AGAIN

The concept of equivalence—that two sets are equivalent if they have the same number of members—does not depend on the sets being finite. So the set of positive integers and the set of negative integers are infinite but equivalent sets, since there is a negative integer to match every positive integer. As Cantor quickly realized (and others, including Galileo, had realized before him), each natural number can be squared, so there is an infinite set of natural numbers and an infinite set of their squares. Yet the squares are a subset of the set of natural numbers. Galileo's conclusion in 1638 had been that the concepts of "equal to," "greater than," and "less than" did not apply to infinity. But Cantor developed instead a concept of transfinite numbers that recognized different sizes of infinity.

AXIOMS AGAIN

Attempts to deal with paradoxes at the heart of set theory led to the development of axiomatic set theory. This aims to develop axioms to form the ground rules for set theory, much as Euclid's axioms form the ground rules for trigonometry. Several conflicting sets of axioms have been proposed, too complex to go into here. The basic criteria for axioms are that they should be

- consistent: it should not be possible to prove a statement and its opposite
- plausible: they should accord with general beliefs about sets
- capable of producing results of Cantorian set theory.

Axiomatic set theory is further divorced from the real world than Cantorian (or "naive") set theory since it requires no knowledge of what the sets discussed are. It concentrates only on relations between sets and their properties in a rather nebulous way that gives fuel to the few mathematicians who still claim set theory deals with nonexistent fictions.

Set theory has influenced many areas of 20th-century mathematics, but remains in turmoil. The search for acceptable axioms recalls the difficulties faced by geometers trying to find new models and rules for non-Euclidean geometries, but so far is a long way off resolution.

Getting Fuzzy

At the heart of set theory is the apparently simple rule that an object either is or is not a member of a set. Aesop's fable of the tortoise and the hare, like Zeno's paradox of Achilles and the tortoise, pits a fast and slow contestant in a race against each other.

Both the hare and the tortoise are members of the set "animals." The hare is a member of the subset "mammals"; the tortoise is a member of the subset "reptiles." Now suppose we had a set of "fast animals." We might say the hare is a member and the tortoise is not. But this is a subjective judgment and proves more difficult with some other animals. What about a dog—is it a fast animal? Or a snake? Or a giraffe? We are likely to say that they are somewhat fast, or quite fast. However, set membership is binary—either an animal is fast or it is not. This is clearly problematic as it requires an absolute cut-off point that is not necessarily satisfactory. If we say that an animal that can run at 15 miles an hour is fast, then an animal that can run at 14.95 miles an hour is not fast; the distinction begins to look silly.

CATERING FOR IMPRECISION

Aristotle identified the problem of the "excluded middle"—the objects that can be classified as neither one thing nor another (a slightly fast animal, for example). But mathematics had no way of dealing with the indeterminate, and the middle ground remained excluded until the 20th century. Bertrand Russell, in his paradoxes of the barber who may or may not shave himself and the set that contains all sets, highlighted the problem again, showing it as a contradiction in set theory.

In the 1920s, the Polish logician Jan Lukasiewicz worked out the principles of multivalued logic, in which statements can take a fractional truth value between 1 (wholly true) and 0 (wholly false). In 1937, philosopher Max Black applied multivalued logic to sets of objects and drew the first "fuzzy" set curves; he called these sets "vague."

From these outlines, the American mathematician Lotfi Zadeh developed the concept of fuzzy logic and fuzzy sets in 1965. These provide a way of working with imprecise values and categories. There is some disagreement about the validity and nature of fuzzy theory. Some mathematicians see it as a variation on probability theory, which can be called possibility theory; others see probability as a special case of possibility in which certainty can be applied.

FUZZY COUNTING

The distinction we saw early on between counting and measuring addresses the problem of things that do not fall wholly into one set or another. Instead of an element belonging to a set or not belonging to it—a binary distinction, with values of either 0 (not a member) or 1 (a member), fuzzy sets can support degrees of membership. Membership of a set can have a value between (and including) 0 and 1.

So in a set of fast animals, a cheetah may have a membership value of 1, Achilles a membership value of 0.5, and a tortoise a membership value of 0.1. Something that does not move at all, like a mature barnacle, would have a value of 0 and not be a member of the set.

Fuzzy theory makes use of linguistic categories, such as "somewhat," "quite,"

and "very." So an animal might be very fast, or quite fast. If 0.6 membership of the set of fast animals is called "quite fast," 0.8 membership might be "very fast." Fuzziness is not about uncertainty, but about the vague boundaries between categories.

Fuzzy sets may overlap. So an animal might have 0.2 membership of the set "fast animals" and 0.8 membership of the set "slow animals." By combining values from more than one set, useful information can be gained that gives a better description of a situation or object than the straightforward binary membership/nonmembership of a conventional set.

Some, but not all, of the mathematics of conventional sets apply to fuzzy sets. In fuzzy sets, an object may be a member of two complementary sets (such as slow animals and fast animals), whereas in conventional set theory this is not possible. The only restriction is that its total membership value for the two sets adds up to 1 (such as 0.2 fast and 0.8 slow).

USING FUZZINESS

Fuzzy logic is the application of fuzzy sets to decision making and computer programs. It is used in many engineering control systems to approximate human judgment and make the operation of a device adapt to prevailing conditions. It is commonly used in consumer electronics, household appliances, and vehicles. A digital camera, for example, uses sensors to determine the light levels and the objects in the view that the photographer is likely to want to focus on (from detecting the edges of objects), then adjusts focus and exposure appropriately. A washing machine determines the best features for the wash cycle from the quantity of washing and how dirty it is, for example. It will calculate optimum amounts of soap and water, the best temperature, and the length of wash required. The first system controlled by fuzzy logic was created by Ebrahim Mamdani and Seto Assilian at Queen Mary College, London in the early 1970s. They wrote a set of heuristic rules for controlling the operation of a small steam engine and boiler, then used fuzzy sets to convert the rules into an algorithm to control the system. The first commercial use of a fuzzy system was to control a cement factory in Copenhagen, Denmark in 1980. Exploration and uses of fuzzy logic increased massively in the 1980s, especially in Japan.

Fuzzy logic is used not only in control but also in expert systems, artificial intelligence, and applications such as voice recognition and image processing software. It tries to minimize the human intervention needed in a system by approximating human judgment. For this it needs an expert human to set up the rules upon which judgments are based, but intelligent systems can then improve themselves by learning from adjustments an operator makes to the settings that the system chooses. In diagnostic medicine, for example, a fuzzy system can look at all the symptoms reported or monitored in a patient and assess the likelihood of different diagnoses based on the degree to which each symptom is present. A confirmed or refuted diagnosis can then be fed back into the system to improve its future performance.

> **SENDAI SUBWAY**
>
> In 1988 Hitachi produced a fuzzy logic system to run subway trains in Sendai, Japan. The trains need only a conductor and no driver. The fuzzy logic system controls acceleration, cruise speed and braking, taking into account safety, comfort, fuel efficiency and the need to stop accurately at target positions (station platforms).

Sets—both fuzzy and classical—have redefined mathematics for the 20th and 21st centuries. In some ways, they have allowed mathematics to be divorced from the real world. Higher set theory deals not with numbers or objects in the world, but with concepts and relations between concepts. Yet in accommodating the imprecision and contingency of the real world, set theory, like fractals, acknowledges the "roughness" of the real world and provides a more accurate (if messier) model of reality than earlier mathematics.

MOVING ON

Set theory works with mathematics far divorced from numbers. As it does so, it becomes increasingly dependent upon logic. Although it may seem that logic has been at the heart of mathematics from the start—after all, Euclid attempted to derive all geometry through a sequence of logical steps—the application of logic was neither rigorous nor closely examined until the 19th century. Set theory, it turned out, could be applied to developing the logic needed to give mathematics firm foundations.

CHAPTER 9

PROVING IT

As in law, everything in mathematics must be proven before it is accepted as true. Even the most blatantly obvious "facts" are not accepted as facts unless a mathematician can provide a rigorous proof. It is not enough to put one apple with another apple to show that one plus one equals two: it must be proven that one plus one always equals two, that there are no cases in which one and one might make one, or zero, or three, or 1.7453.

Often, it is much harder to prove something than to discover it and decide that it is almost certainly true. Sometimes, it takes many centuries for a theorem to be proved, as in the case of Fermat's Last Theorem. But it is the proof that defines a theorem—it must be possible to demonstrate its truth through a line of logical reasoning from axioms and other established theorems.

That the sun has always risen is no proof it will rise tomorrow.

Problems and Proofs

It took Jacob Bernoulli 20 years to prove that tossing a coin a large number of times will give close to a 50:50 split between heads and tails—yet as he pointed out, the result is obvious to anyone. Why did he bother? And why did it take so long?

Although the Ancient Egyptians and Babylonians were content to work with specific examples and problems, the Greeks moved toward theorems and axioms that could be applied universally—they demanded proof. Proving that an idea holds true requires some kind of logical theoretical treatment, since it is not possible to try out all possible cases—to test Pythagoras' theorem for all possible right-angled triangles, for example.

Proofs aim to find fruitful relationships between mathematical statements and objects. For this reason, even theorems that have been adequately proven in the past—such as Pythagoras' theorem—may be proven anew, opening up fresh avenues for exploration. Over time, simpler proofs are discovered and the earlier, often cumbersome, proof can be replaced.

Many developments in mathematics came about as the result of people testing and trying to prove theorems and axioms and even doubting long-held beliefs. The dispute over Euclid's fifth postulate, for instance, was the spur to much of the

> "Each problem that I solved became a rule that served afterward to solve other problems."
>
> René Descartes

progress made in geometry and ultimately to the emergence of new, non-Euclidean geometries in the 19th century.

Rigor in mathematical proof increased at the end of the 19th century when mathematics and logic came together. A systematic notation for logic came to be used by mathematicians and some philosophers. The development of set theory required a method of representing logical relationships and a way of dealing with concepts that did not necessarily involve any numbers at all. Set theory even became a useful means of demonstrating mathematical theorems.

UNBELIEVABLE PROOFS

A famous problem that produces a proof that many people find hard to accept is the Monty Hall paradox. Named after the host of a US game show, it goes like this:

Suppose a game show host shows you three doors. Behind two of them there is a goat; behind the last there is a car. The host invites you to pick a door. He will then open another door, revealing a goat, and give you the chance to change your choice. Will your chances of winning be improved if you switch doors? (The problem assumes that you would rather have a car than a goat.)

> "Nobody blames a mathematician if the first proof of a new theorem is clumsy."
>
> Paul Erdös

Most people say their chances of getting a car are unaffected if they switch doors. Mathematicians say that the chance of getting the car is increased if you switch doors: you had a 1 in 3 chance of choosing a car, and this is unaffected by the opening of another door; the chance that you chose correctly is still 1 in 3. If you switch, you are making a new choice, where the chances are 2 in 3. Switching will get you a car 67 percent of the time, but staying with the first door will yield a car only 33 percent of the time. The logic is easier to follow if you think of 1,000 doors with goats behind 999 of them. Your chances of picking the door with the car the first time are 1 in 1,000. After 998 goats have been released to run amok, there is a 1 in 2 chance that the other door hides the car.

The obvious objection here is that there must also be a 1 in 2 chance that the original door hides the car, since probabilities must add up to one. The trick is that the problem is not as it appears. Your choice is random, but the host *knows* where the car is. If the host randomly opened doors, coincidentally picking those that concealed goats, the chance of finding a car at the end would be the same as the chance of finding a goat, whether or not you switched doors.

The proof of this problem uses mathematical notation to show probabilities and breaks it down into small, logical steps, which naturally follow one from another. This is how mathematicians now demonstrate truth. But it has not always been the case.

EARLY PROOFS

The earliest known mathematical proofs are said to have been provided by Thales.

Tradition maintains that Thales proved that the angles at the base of an isosceles triangle are equal, that a diameter cuts a circle into two equal parts, that opposite angles formed by two intersecting lines are equal, and that two triangles are identical if any two angles and one side are equal. Since none of Thales' writings survives, it is impossible to say whether he really produced rigorous proofs of these theorems. Around 50 years later, Pythagoras may have proved the theorem for right-angled triangles that bears his name.

Since the time of Thales and Pythagoras, the basis of proof in mathematics has been to derive more complicated statements from facts that are apparently simpler (though they may not actually be simpler). Generally, anything in geometry that can be demonstrated in clear, logical steps from Euclid's postulates counts as proven, for instance. But this does not mean that a new idea is deduced first from the existing facts. Mathematicians commonly have the idea first—perhaps as

---

DEDUCTIVE PROOF THAT 1 = 2

Let $a = b$. So it follows that

$$a^2 = ab$$
$$a^2 + a^2 = a^2 + ab$$
$$2a^2 = a^2 + ab$$
$$2a^2 - 2ab = a^2 + ab - 2ab$$
$$2a^2 - 2ab = a^2 - ab$$

This can be rewritten as

$$2(a^2 - ab) = 1(a^2 - ab)$$

Dividing both sides by $a^2 - ab$ gives

$$2 = 1$$

---

an intuition, or as something suggested by the results of an experiment or an exploration—and then turn to the known facts to prove it. Sometimes, an attempt to find proof refutes the new theory and it must be rejected. Sometimes, finding a proof appears an intractable problem and the theorem remains unproven—for hundreds of years in some cases.

PROOF BY DEDUCTION

Proof by deduction works in small steps to deduce new truths from known truths. For example, if we say, "Humans are mammals" and "Peter is a human," we can then say, "Peter is a mammal." Deduction is not wholly reliable, even if the initial statements are genuinely true, as the reasoning may not be valid. So we might say, "Humans are mammals" and "Peter is a mammal," therefore "Peter is a human"—but the first statements would also be true if Peter were a dog or a hamster, or any other mammal. Proof by deduction was used extensively by the Ancient Greeks and by medieval mathematicians. Parmenides is credited with the first proof by deduction in the fifth century BCE. Modern mathematicians accept proof by deduction as long as it is sufficiently rigorous.

INDIRECT PROOF

Another method of proof, which also originated with the Greeks but was refined and defined more rigorously much later, is indirect proof. There are several types of indirect proof, including proof by contradiction and proof by *reductio ad absurdum*. Proof by contradiction aims to prove a statement is true by showing that its opposite is not true. Proof by *reductio ad absurdum* aims to prove a statement is true by using it to prove untrue something that is known to be true (so producing an absurd result). Hipassus' proof of the existence of irrational numbers was an indirect proof and is the earliest known.

PROOF BY INDUCTION

The Greek model of proof was followed by the Arab mathematicians and taken over from them in the Middle Ages by early European scholars. But in 1575 a new

ALL HORSES ARE THE SAME COLOR

The Hungarian mathematician George Pólya (1887–1985) used proof by induction to show that all horses are the same color. The case for n = 1 (one horse) is clear—a horse can only be the same color as itself. Now assume the theory is correct for n = m horses. We have a set of m horses, all the same color (1, 2, 3, … m). There is a second set of (m + 1) horses (1, 2, 3, … m + 1). We take out one horse from this last set, so that it contains horses (2, 3, … m + 1). The two sets overlap; this second set is a set of m horses, which we know is a set of horses the same color. By the principle of induction we can continue this for all further horses, therefore all horses are the same color.

The argument is, of course, invalid as the statement is not true. The crucial point is that when n = 2 the statement does not hold true: for this value, the sets do not overlap (the first contains only horse 1, the second contains only horse 2).

DAVID HILBERT (1862–1943)

David Hilbert is considered one of the most influential mathematicians of the 19th and 20th centuries. He was born in East Prussia in an area that is now part of Russia. As a student, he met Hermann Minkowski and the two stayed lifelong friends, cross-fertilizing each other's mathematical ideas.

Hilbert worked in many fields, but is best known for his contributions to the axiomatization of mathematics.

He began as a pure mathematician and, when he turned his attention to physics around 1912, was horrified at what he considered the sloppy approach to math taken by most physicists. Hilbert also devised a conceptual space that had infinite dimensions (called a Hilbert space). He and his students contributed to the math behind Einstein's Theory of Relativity and quantum mechanics.

model emerged; in *Arithmeticorum Libri Duo*, Francesco Maurolico (1494–1575) gave an early description of mathematical induction. Aspects of the method can be found in earlier works by Bhaskara and al-Karaji (*ca.*1000 CE). Proof by induction was also developed independently by Levi ben Gerson (1288–1344), Jacob Bernoulli, Blaise Pascal, and Pierre de Fermat.

Proof by induction works by showing firstly that a hypothesis holds true for a first value (often n = 1), then that if it holds true for any later value (n = m), it must also be true for n = m + 1. Since it holds true for n = 1, it therefore holds for n = 2, for n = 3, and so on—that is, for all values of n.

It's a bit like a row of dominoes, arranged on end and equally spaced so that if one falls it will knock the next over. If knocking the first domino over causes the next to fall, it will inevitably follow that they will all fall.

Maurolico used proof by induction to demonstrate that the sum of the first n odd integers is n^2:

$$1 + 3 + 5 + 7 + 9 + \ldots + (2n - 1) = n^2$$

ASKING QUESTIONS

With the advent of calculus, complex numbers and later non-Euclidean geometries, more and more was demanded of proof.

Berkeley's objection to calculus as dealing with the "ghosts of quantities" was a spur to greater rigor, not only in defining the quantities and concepts with which mathematicians were working but in providing proofs.

Greeks. Plato, the earliest rigorous writer on logic, died in 347 or 348 BCE.

Plato presents his philosophical works in the form of dialogues, or conversations, between philosophers. They read as arguments, with each participant putting forward his case in a series of statements which his opponent then refutes and he then defends. The argument becomes increasingly complex as the subject is tackled rigorously. This method, called dialectic, formed the model for logical debate until the Middle Ages. Although logic was a major concern of these medieval scholars, they did not think to apply it to mathematics. It took more than 2,000 years for logic and mathematics to come together properly.

But it was the 19th century that saw the great revolution in mathematical proof as new methods of logic were developed and people for the first time tried to apply formal logic to mathematics. This required a reassessment of the very basis of mathematics and brought mathematics and philosophy together. Mathematicians, unsettled by recent discoveries that threw long-accepted truths into doubt, sought new proofs and questioned even the most fundamental ideas underpinning their discipline. Suddenly, nothing could be taken for granted.

MATHEMATICS BECOMES LOGICAL

One of the first to tackle the issue was the Italian mathematician Giuseppe Peano (1858–1932). He wanted to develop the whole of mathematics from fundamental propositions using formal logic. He developed a logic notation, but also a hybrid international language which he hoped would be used for scholarship. Called Interlingua, it was based on the vocabulary of Latin, French, German, and English, but used a very simple grammar. His use of it hindered the acceptance of his mathematical work.

Being Logical

At the end of the 19th century and start of the 20th century there was a flurry of interest in the application of logic to mathematics or, more precisely, the derivation of mathematics from logic. It came about largely as a result of rapid changes in mathematics and its applications, and criticisms of its rigor and validity.

Proof in mathematics is only part of the larger topic of logic which has developed and grown since the time of the Ancient

The breakthrough in relating mathematics to logic came with the work of the German logician and mathematician Gottlob Frege (1848–1925), who has sometimes been called the greatest logician since Aristotle. He set out to prove that all arithmetic could be derived logically from a set of basic axioms and he is essentially the founder of mathematical logic. He devised a way of representing logic using variables and functions.

A SEARCH FOR NEW AXIOMS

The German mathematician David Hilbert laid the foundations for the formalist movement that grew up in the 20th century by requiring that all mathematics should depend on fundamental axioms from which everything else can be proven. He required any system to be both complete and consistent, incapable of throwing up any contradictions from the application of its axioms. He reformulated Euclid's axioms himself as the first step in trying to find this faultless axiomatic basis for math. Hilbert famously proposed 23 problems which were still to be solved in 1900. These effectively set out the agenda for 20th century mathematicians.

The most important of Hilbert's problems for the development of logic in mathematics is the second. He proposes that it is necessary to set up a system of axioms "that contains an exact and complete description" of the relations between basic ideas and requires "that they are not contradictory, that is, that a definite number of logical steps based upon them can never lead to contradictory results." In particular, this was seen as a call for axioms to prove the basics of Peano arithmetic.

Answering Hilbert's call for an axiomatic basis for all mathematics, Bertrand Russell and Alfred North Whitehead published the three-volume *Principia Mathematica* in 1910–13. Ambitiously named after Newton's seminal work with the same title, the book aims to derive all of mathematics from a set of basic axioms using the symbolic logic set forth by Frege. It covers only set theory, cardinal numbers, ordinal numbers, and real numbers. A planned volume to cover geometry was abandoned as the authors were tired of the work. After getting a good way into the work, Russell discovered that a lot of the same ground had been covered by Frege and he added an appendix pointing out the differences and acknowledging Frege's prior publication.

The test of the *Principia* rested on whether it was complete and consistent in Hilbert's terms—could a mathematical statement be found that could not be proven or disproven by *Principia*'s methods, and could any contradictions be produced using its axioms?

MOVING THE GOALPOSTS

Before *Principia* had a chance to stand the test of time, the key questions were taken away by German mathematician Kurt Gödel. He produced two "incompleteness theorems" (1931) which dealt with Hilbert's proposal for the axiomatization of mathematics.

The first stated that there could be no complete and consistent set of axioms, since for every sufficiently powerful logical system there is always a statement G that essentially reads, "The statement G cannot be proved." If G is provable, then it is false and the system is inconsistent. If G is not provable, then it is true and the system is incomplete. The second theorem states that basic arithmetic cannot be used to prove its own statements and nor, by extension, can it be used to prove anything more complex.

Logic and Computers

During the 20th century, the development of computers has given logic and mathematics a field of their own. Computer programs use logical sequences to carry out calculations. This is the basis of all computer applications, even those that look nothing like mathematics to the user, such as animation, music production, and image processing. Computers can also be used to test theorems. They can produce a proof by exhaustion—which involves trying all possible values—which a human could not manage. There are, too, computer programs to construct proofs by other methods.

The program Vampire, developed at Manchester University by Andrei Voronkov, has won the "world cup for theorem provers" six times (1999, 2001–05). Perhaps the time has come when computers, with their impeccable logic, will take over from human mathematicians as the experts at applying logic to mathematics, or tracing it in mathematics.

What Were We Talking About?

At no point in this book have we stopped to ask what mathematics *is* or if, indeed, it "is" anything. This might seem like a considerable oversight. But on the whole mathematics crept up on humanity, made

itself at home with no introduction, and encouraged us to build our cultural edifice

> "For any consistent formal, computably enumerable theory that proves basic arithmetical truths, an arithmetical statement that is true, but not provable in the theory, can be constructed. That is, any effectively generated theory capable of expressing elementary arithmetic cannot be both consistent and complete."
>
> Kurt Gödel, 1931

around it. There was no point at which it was sensible to ask what it was all about.

At the start of the 20th century, mathematics turned to fundamental questions about its very nature. A central question can be briefly summarized as "Is mathematics discovered or invented?" There are three principal positions. A Platonist realist view, such as that of Kurt Gödel, says that the laws of mathematics are everywhere true and immutable, like the laws of nature. Mathematicians discover them; they are pre-existing. A formalist view, such as that taken by David Hilbert, says that mathematics is a codification, a language, or even a game in which theorems are built on axioms through logical demonstration. There is no particular reason to prefer one set of axioms over another if both sets seem to hold true. This view was dealt a near-fatal blow by Gödel's incompleteness theorems, which showed that no set of axioms could be entirely complete and consistent. Finally, the intuitionist view holds that mathematics is entirely a fabrication of the human mind, constructed to explain the world we find around us but having no existence or validity outside human culture. This view was propounded by the Dutch mathematician L. E. J. Brouwer (1881–1966) and he was remorselessly ridiculed and persecuted for it (not least by Hilbert).

Over the last hundred years, the question of the foundation of mathematics has not been answered, but has slipped out of view. Hilbert's formalist stance suffered from the assault of the incompleteness theory, yet logic and axioms still lie at the heart of mathematics as it is practiced. A more modern view is the empiricist one promoted by W. V. Quine (1908–2000) and Hilary Putnam (born 1926). They maintain that the existence of numbers and other mathematical entities can be deduced from observation of the real world. It is related to realism, but is more grounded in reality and human culture.

Quine's view is that mathematics seems to be "true" because all our experience and science is woven around it and appears to endorse it. It would be very difficult to rebuild our model of the universe without mathematics.

If the last sentence sounds like a challenge, it is one that was taken up by American philosopher Hartry Field (born 1946). In the 1980s he proposed that mathematical statements are all fictional and that science can be created without mathematics.

According to his fictionalism doctrine, mathematical statements are useful structuring devices, but should not be accepted as literally true. And why would we "make up" mathematics? One answer is that the structure of the human mind makes it inevitable. The embodied minds theory is based in cognitive psychology; it was developed for mathematics by American cognitive linguistic George Lakoff and psychologist Rafael Núñez. Their argument, expressed in their book *Where Mathematics Comes From* (2000), is that the structure of the human brain and the way our bodies operate in the world has dictated the way we have developed mathematics. As we can't divorce ourselves from our brains to examine the universe without our cognitive processes getting in the way, we will not be able to tell whether mathematics

"Mathematics may or may not be out there in the world, but there's no way that we scientifically could possibly tell."
George Lakoff, 2001

has any existence outside human culture.

Plenty of mathematicians disagree with Lakoff and Núñez, and with the proponents of all the other ideas outlined here, and they will no doubt argue the case for many decades or centuries to come. The question of its foundations has little impact on our day-to-day use of mathematics. We will carry on playing the lottery, building aircraft, looking for life in outer space, and insuring against catastrophes without knowing whether mathematics is in any sense "real" and "out there," just as the Egyptians built their pyramids and the Incas counted their llamas without giving the matter a second thought.

GLOSSARY

algebraic geometry Geometry using algebraic equations and expressions.

algorist Someone who calculates using the Hindu-Arabic number system to carry out arithmetic rather than using an abacus.

algorithm A rule for carrying out a calculation.

analytic geometry Geometry using coordinates.

axiom A basic law that is self-evident and requires no proof.

base The basis of a counting system; the base number is that to which numbers are counted before shifting the place value (to tens, hundreds, etc).

binary Counting system that has only two digits, 1 and 0 (base 2).

binomial coefficients The sequence of coefficients used with variables when a binomial expression is expanded.

calculus The branch of mathematics concerned with calculating the sum of infinitesimal quantities to approximate the area under a curve or the rate of change of a curve.

chord A straight line joining the ends of an arc (a portion of the circumference of a circle).

coefficient A constant or number by which a variable is multiplied in an algebraic expression.

commensurable (Or more than one quantity) able to be measured or compared to a conic section—a curve produced by cutting a section through a cone.

conics The family of curves produced by cutting through a cone, or the study of these curves.

conjecture An unproven theorem.

cosine The ratio of the side adjacent to an angle to the hypotenuse in a right-angled triangle.

cotangent The ratio of the side adjacent to an angle to the side opposite the angle in a right-angled triangle.

decimal fraction A number in which fractional parts are expressed as a decimal (showing tenths, hundredths, thousandths, and so on).

differential calculus (differentiation) Method for calculating the slope of a curve at a particular point.

Diophantine equation An equation in which all the numbers involved are whole numbers.

fractal A curve or other figure that repeats its overall pattern or shape in portions of constantly reducing size, so that a portion of the figure when magnified looks the same as the whole figure.

function A mathematical expression with one or more variables.

hyperbola A curve produced by slicing through a cone with a plane with a smaller angle at its axis than the side of the cone.

hyperbolic geometry Geometry that deals with shapes drawn on curved surfaces.

imaginary number A number that involves the square root of -1.

incommensurable (Of more than one quantity) not able to be directly compared or measured by the same standard.

infinitesimal Very small quantity, tending toward zero.

integral The product of integration.

integral calculus (integration) Method of calculating the area under a curve by approximating the sum of a large number of infinitely thin slices of the area.

irrational number A number that cannot be expressed as the ratio of two whole numbers.

limit The lowest or highest value to which a function will be calculated.

logarithm The power to which a base figure (usually 10 or e) must be raised to give a specified number.

optics The study of lenses, vision, and light.

parabola Curve produced by slicing through a cone with a plane parallel to the side of the cone.

parallel postulate Euclid's fifth postulate, which states the condition that must be met for lines not to be parallel (and so by reversing the postulate gives the condition for lines to be parallel)

perpendicular At right angles to.

perspective geometry The study of how three-dimensional figures appear and can be represented in two dimensions.

polynomial equation An equation that involves non-zero powers of a variable (e.g. $x2 + 4x + 1 = 0$) .

quadratic equation An equation of the form $ax2 + bx + c = 0$

quinary Relating to the number 5.

rational number A number that can be expressed as a ratio of two whole numbers.

real number Any positive or negative number that does not involve the square root of -1.

Riemann geometries Geometries of surfaces that do not follow the standard rules of traditional planar geometry.

set Related group of entities.

sexagesimal Relating to the number 60.

sine The ratio of the side opposite an angle to the hypotenuse in a right-angled triangle.

spherical trigonometry The study of triangles drawn on the surface of a sphere.

tangent The ratio of the side opposite an angle to the side adjacent to the angle in a right-angled triangle.

theorem Statement of a rule that is not self-evident but which can be proven by logical steps.

topology The study of geometric properties that are not affected by changes of shape or size.

transfinite numbers Numbers that relate to infinities of different magnitudes; so the infinite number of whole numbers is smaller than the infinite number of real numbers.

triangulation Procedure for measuring or mapping a surface by dividing it into triangles and calculating distances and angles.

For Further Reading

Acheson, David. *1089 and All That: A Journey into Mathematics*. New York, NY: Oxford University Press, 2002.

Gowers, Timothy. *Mathematics: A Very Short Introduction*. New York, NY: Oxford University Press, 2002.

Pickover, Clifford A. *The Math Book: From Pythagoras to the 57th Dimension, 250 Milestones in the History of Mathematics*. New York, NY: Sterling, 2009.

Web Sites

Due to the changing nature of Internet links, Rosen Publishing has developed an online list of Web sites related to the subject of this book. This site is updated regularly. Please use this link to access the list:

http://www.rosenlinks.com/hos/math

INDEX

206